Welding For Dummies®

by Steven Robert Farnsworth

Wiley Publishing, Inc.

Welding For Dummies®

Published by
Wiley Publishing, Inc.
111 River St.
Hoboken, NJ 07030-5774
www.wiley.com

For general information on our other products and services, please contact our Customer Care Department within the U.S. at 877-762-2974, outside the U.S. at 317-572-3993, or fax 317-572-4002.

For technical support, please visit www.wiley.com/techsupport.

Wiley also publishes its books in a variety of electronic formats. Some content that appears in print may not be available in electronic books.

Library of Congress Control Number: 2010933466

ISBN: 978-0-470-45596-8

Manufactured in the United States of America

10 9 8 7 6 5 4 3 2 1

About the Author

Steven Robert Farnsworth hails from Iowa, close to the age of dirt being born in 1955. He attended Archer School until they closed it and went on to Sanborn Community High School, where he graduated in 1973. He was one of the lucky ones. While attending high school, he also attended a vocational school, graduating with a welding diploma.

Steve enlisted in the U.S. Navy in July of 1973. After boot camp, he attended HTA (Hull Maintenance Technician) in San Diego, California, and transferred to the USS ARD 30 (the floating dry dock for fast attack submarines), where he worked in the repair division. Steve was then sent to C1 Welding School in San Diego, where he qualified as a high pressure plate and high pressure pipe welder. After attending C1 Welding School, he was transferred to the USS Basilone DD824 until his honorable discharge in July of 1977. Steve began his civilian career at a construction company in Spencer, Iowa, working as the welder, semi driver, and heavy equipment operator. In August 1979, he acquired the position of Welding Instructor at Iowa Lakes Community College, teaching the following classes:

- Oxyacetylene Theory and Lab
- Electric Arc Theory and Lab
- Structural Welding
- Brazing and Soldering
- Mig and Tig Theory
- Mig and Tig Lab
- Pipe Welding
- Production Welding
- Special Processes

In 1984 Steve left teaching and went back into the Navy, receiving orders to the USS White Plains AFS-4 (the Orient Express), home ported in Guam. After 36 months, he received orders to the USS Hunley AS-31, home ported out of Norfolk, Virginia. After four years in that naval tour, he returned to instructing at Iowa Lakes Community College and has been there ever since. Steve is a Certified Welding Educator and Certified Welding Inspector through the American Welding Society.

Dedication

I dedicate this book to all the welders and welding students who put everything together that makes this world go around. I cannot imagine a world without welded products. So thank you, welders who have breathed a little smoke and saw the flash of products being welded together. I would also like to thank Iowa Lakes Community College for allowing me to write this book and putting up with me for all these years.

Author's Acknowledgments

When I was young, my first job was moving chicken manure for 90 cents an hour. (Believe me, you never had a cold after you scooped that stuff.) Welding at the local factory paid 35 cents more an hour. (Wouldn't you change jobs?) At that time, who would have ever thought I would author a book on something that I enjoy doing? I would like to thank all the people who have helped me teach welding students how to weld over the last 25+ years. I hope this welding book helps you become the welder that you want to be.

Publisher's Acknowledgments

We're proud of this book; please send us your comments at http://dummies.custhelp.com. For other comments, please contact our Customer Care Department within the U.S. at 877-762-2974, outside the U.S. at 317-572-3993, or fax 317-572-4002.

Some of the people who helped bring this book to market include the following:

Acquisitions, Editorial, and Media Development

Project Editor: Natalie Faye Harris

Acquisitions Editor: Michael Lewis

Copy Editor: Megan Knoll

Assistant Editor: David Lutton

Technical Editor: Emily Gottsche

Editorial Manager: Christine Meloy Beck

Editorial Assistant: Rachelle Amick

Art Coordinator: Alicia B. South

Cover Photos: © iStockphoto.com/Fertnig

Cartoons: Rich Tennant (www.the5thwave.com)

Composition Services

Project Coordinator: Sheree Montgomery

Layout and Graphics: Ashley Chamberlain, Joyce Haughey, Mark Pinto, Christine Williams

Photography: Christina Kjar

Proofreaders: John Greenough, Penny Stuart

Indexer: Steve Rath

Publishing and Editorial for Consumer Dummies

Diane Graves Steele, Vice President and Publisher, Consumer Dummies

Kristin Ferguson-Wagstaffe, Product Development Director, Consumer Dummies

Ensley Eikenburg, Associate Publisher, Travel

Kelly Regan, Editorial Director, Travel

Publishing for Technology Dummies

Andy Cummings, Vice President and Publisher, Dummies Technology/General User

Composition Services

Debbie Stailey, Director of Composition Services

Contents at a Glance

Table of Contents

Introduction

Welding has become one of the most important trades in the world, and that isn't likely to change anytime in the near future. So many of the objects people have and need are created either directly or indirectly by welding. If everyone woke up one morning and no one could remember how to join metals, the world would be a very different place by the afternoon.

But welding isn't just important — it's also fun. The idea of welding as a hobby is catching on more and more. It's an extremely versatile skill that can be quite rewarding after you get the hang of it. Something is very empowering about knowing that you can harness some pretty powerful forces — electricity and intense heat — to melt metals and join them together. Even experienced welders get a kick out of the fact that they can take a machine and a few pieces of metal and create something new, functional, and even beautiful. That's an extremely fulfilling feeling, and I think it's a product of welding that people don't always mention when they talk about the trade.

One quality of welding that people *do* talk about a lot is its usefulness. You can use welding skills to accomplish a lot, whether you want to eventually make a career out of welding or just have the ability to make and fix metal objects for your personal pursuits. Over the years I've taught and worked with both kinds of welders, and I know that after they really figured out the ins and outs of welding, they were able to do things that made their personal and professional lives a lot easier.

About This Book

Welding For Dummies helps you understand the basics of how welding works and lets you begin practicing several of the most prominent and useful welding techniques. I walk you through the fundamentals that hold true for all types of welding, and I dig into the details of specific welding processes — stick, mig, tig, and more — to show you how to practice those skills in a safe, productive way. Don't worry; I don't have you welding the Statue of Liberty's torch back onto her hand or anything, but I do hope this book puts you well on your way to achieving the welding goals you've set for yourself.

One of my favorite aspects of *Welding For Dummies* is that you can move around within it however you want and still end up with a huge amount of welding knowledge. You may initially be interested in one welding process but quickly discover you should be reading about a totally different process, and that's okay — you can jump to that other discussion without worrying that you've missed something important. Just beware of paper cuts from flipping back and forth between chapters.

Conventions Used in This Book

Here are a few conventions I use to make reading this book even easier:

- The world of welding is full of jargon, so I present new terminology in *italics* and make sure to give a definition nearby.
- **Bold** text highlights the action parts of numbered steps and also designates keywords in bulleted lists.
- I've tried to stick to welding standards supported by the American Welding Society (AWS), which is the largest and most prominent welding organization in the United States.
- All Web addresses appear in `monofont`. When this book was printed, some Web addresses may have needed to break across two lines of text. If that happened, rest assured that the address doesn't contain any extra characters (such as hyphens) to indicate the break. So when using one of these Web addresses, just type in exactly what you see in this book, pretending as though the line break doesn't exist.

What You're Not To Read

Far be it from me to tell you what you should read, but allow me to make one quick point. In several spots throughout this book, I include *sidebars* (gray shaded boxes) that contain interesting (and possibly entertaining, depending on what kind of mood you're in) information that you don't absolutely have to read in order to understand and practice welding. If the how-to, functional information in the book is the entrée, the sidebars are like garnish. Not parsley, though — I like to think that the sidebars are at least a little more interesting and useful than an herb that tastes funny and doesn't do much more than crowd a plate. You can also skip anything with a Technical Stuff icon; this information is more technically involved than the basics you need to weld.

Foolish Assumptions

I'm not really crazy about guesswork, but I did make a few assumptions about you as I wrote this book. (They're all nice, I promise.) If any of the following statements applies to you, this book is for you.

- You've never welded but want to know more about metals and how to join them by using welding.
- You've welded a little but really want to figure out how to improve and start taking advantage of all welding has to offer.
- You've done a fair amount of one type of welding but want to expand your skill set so you can weld with a variety of different processes and techniques.
- You understand a few basic tools (such as hammers and screwdrivers) and what they do.
- You know how important taking necessary safety precautions is to keep yourself (and others) out of harm's way.

How This Book Is Organized

This book is divided into six parts. Each part offers something different, but all of them are geared toward helping you figure out welding processes and put them to good use. Here's a quick look at what you can find in each part.

Part I: Understanding Welding Basics

This part provides the kind of welding information that crosses all types of welding. If you're really just starting out in welding, this part is a good first stop for you because it gives you the lowdown on metals (especially the ones that are commonly used in welding), the tools and equipment you use for welding, and the kind of environment you need in order to weld successfully. It also includes the chapter that's without a doubt the most important one in the book. That's Chapter 3, and it's all about welding safety.

If you read only one chapter in the book, let it be Chapter 3. Welding is a fantastic skill, but it's not worth getting hurt over.

Part II: Welding on a Budget: Stick and Tig Welding

Part II focuses on stick welding (the most commonly used welding process) and tig welding (also a great, useful technique). You can read all about the advantages and disadvantages of both stick and tig and understand how they work and what makes them unique. I explain the different equipment you need if you want to get into stick or tig welding, and I also give you plenty of information on how you can try out the techniques.

Part III: Discovering Mig Welding

Mig welding is a fast, efficient welding process, and it's great for new welders because it's relatively easy to pick up and get started with. I devote Part III to the basics of understanding and executing mig welding.

Part IV: Getting Fancy: Plasma Cutting, Oxyfuel Cutting, and Other Processes

Arc welding isn't the only way you can weld — welding includes lots of other processes, such as soldering, brazing, and gas welding, that are all useful in their own distinct ways. I cover those processes in Part IV.

I also cover a few cutting processes in this part, because cutting is an important task in any welding shop, and you'll probably need to do some (or a lot) of it if you stick with welding for any extended period of time.

Part V: Putting Welding into Action with Projects and Repairs

This part is probably the most fun because it gives you a chance to try out your welding skills and build some great, useful items. The chapters contain a welding project or two that is designed with the beginning welder in mind. You can read about how to build a portable welding table, a torch cart, a campfire grill, and more! This part also includes a chapter that helps you to figure out whether fixing something or buying (or building) it new makes more sense.

Part VI: The Part of Tens

If you've read a *For Dummies* book before, you already know all about this part. The Part of Tens is always a favorite; it features lists full of useful information in an extremely easy-to-read format. You can read about the advantages to becoming a certified welder, the tools that every welder wants, and more. There's also a glossary to help you with basic welding terminology.

Icons Used in This Book

Throughout the book, you'll occasionally notice little pictures in the margins. These icons help flag specific information I want to highlight; check out the following list for details on what those icons indicate.

When you see this icon, expect to find a helpful bit of information that will help save you time and money and keep you from making mistakes when you're welding.

If I really want you to slow down and commit something to memory, I use this icon. It's important stuff, so take the time to read it!

The last thing I want is for you to get hurt, or for you to hurt others or damage property. With that in mind, please pay attention to these icons so you can keep from hurting someone (yourself included) or damaging your equipment or surroundings.

This icon denotes technical or historical information that's more involved than what you need for your basic welding practice.

Where to Go from Here

I know what you're thinking: With all of this terrific, useful welding information, where do I begin?

I certainly don't want to tell you what aspect of welding you should want to read about first — that's for you to decide — but I do make one request. If you're new to welding, or if you aren't completely familiar with the practices of welding safety, please go directly to Chapter 3 and read up on it. You really do need to know how to keep yourself safe as you start or continue your welding experience, and Chapter 3 goes a long way toward keeping you out of harm's way.

After you're done reading Chapter 3, please feel free to jump around in the book however you see fit. There's a whole world of welding out there, and *Welding For Dummies* is a great way for you to start exploring it.

Part I

Understanding Welding Basics

The 5th Wave By Rich Tennant

"This type of welding works best on steel and aluminum. Not so much on Lego blocks."

In this part . . .

Welding isn't the type of skill that you can jump into without any background information, or at least a basic understanding of how it all works. (That's why mother birds push their chicks out of the nest and make them fly instead of pushing them out and making them mig weld.)

With that in mind (the part about welding, not the part about the chicks), in this part I tell you all about metals — specifically, those you're likely to work with as a beginning welder. I also take a full chapter to clue you in on how to set up your welding shop, which may be trickier than you think. (No, you can't just clear the junk out of one corner of your garage and start welding there.) This part also contains the most important chapter in the book: the safety chapter. You can jump around all you want in this book, and read whatever you feel like. But unless you already know all about welding safety (and even if you do), I beg you to read Chapter 3 before you try any sort of welding operation.

Chapter 1

Diving Into the World of Welding

In This Chapter

- Discovering the main uses for welding
- Examining common welding metals
- Paying special attention to welding safety
- Taking a look at welding methods
- Thinking about what's in store for welding in the future

Ever since our early ancestors starting making ornaments out of gold thousands of years ago, metal has played an important role in the lives of all people. Just take a second to look around and think about all the various kinds of metal that are nearby. Dozens (if not hundreds) of metal items are probably all around you, and the items that aren't made out of metal were likely manufactured by using metal equipment.

By and large, metal is tough stuff. (That's one of the reasons why it's so useful, of course.) Throughout history, humans have needed to come up with more and better ways to defy the strength of metals, bending, cutting, and joining it so they can take advantage of its many useful properties. One of the biggest and most important advancements on that front has been the advent and development of welding. Welding allows humans to connect pieces of metal in remarkably strong, sturdy ways, and it has opened up seemingly endless possibilities for what people can do with metallic materials.

This chapter introduces you to all things welding, including its importance, the materials, equipment, and methods you use to accomplish it, and the need for safety precautions while doing it. In addition, the chapter gives you a glimpse into welding's crystal ball.

If You Can't Beat 'Em, Join 'Em: Understanding Why Welding Matters

Welding is the process of using heat to join metals. When you're looking to join metals, you can find no easier or more cost effective way to get the job done than welding — it allows you to join metals in a way that's faster, more versatile, and more dependable than any other process (by a long shot). (And no, using duct tape doesn't count because that's not really fixing anything.) The availability and cost of so many of the items you depend on every day are kept within your reach because of the widespread use of welding processes. Just how prominent is welding? Well, it's estimated that half of the U.S. gross national product is affected by welding. That's about $7 or 8 trillion. How many other skills or trades can claim that much of an impact? Not many.

The uses of welding break down into two very broad categories: fabricating and repairing. The following sections offer a little more detail on both of those divisions.

Fabricating metal products

In welding, *fabricating* simply means that you're taking pieces of metal and welding them together to create something new. That can be as simple as welding a few pieces of metal together at a 90-degree angle to make a pair of bookends in the welding shop you set up in your backyard, or as complex as using underwater arc welding to help build a section of submerged pipeline off the coast of Angola. (Don't worry — you can expect a lot more of the former than the latter in this book!)

Most metals can be joined by one welding process or another, so in theory you don't have many limits when it comes to fabricating. However, for a new welder the amount of fabricating you do with your newfound welding skills is often limited to some degree by cost (some metals can be pretty expensive), time (if you're welding as a hobby, chances are your fabricating time takes a backseat to other obligations like your job and your family), and degree of difficulty. Because developing your welding skills takes time, some fabrication projects may be out of your reach in the short term.

Repairing metal pieces or products

The difference between fabricating and repairing is simple. When you weld to fabricate, you're making something new. When you weld to *repair,* you're welding on something that already exists but needs fixing or modifying.

Repairing can be as simple as welding to fix a tine on your favorite old rake, or welding to fix a crack in a helicopter fitting assembly. (Of course, I lean a lot more toward rake repair than helicopter maintenance in this book!) Although metals are durable and tough, they do break down because of damage or repetitive use, and when that happens, welding is the best way to fix them.

The big question with repair work is whether it makes more sense (especially with regard to time and money) to make a repair or simply replace the broken part or product. That's not always an easy call to make, and I address the various facets of that question in Chapter 18.

When you're welding to repair something, your goal should always be to produce a weld that's stronger than the original piece or product. If you're going to be working on something, why not improve it?

Tracing the history of welding

Welding is one of the newest metal-working trades; it can be traced back to about 1000 B.C. Most historians agree that the first kind of welding done by humans was the lap welding of gold, which was used to create simple gold ornaments. But welding really started to take shape when people figured out how to hammer brass and copper together to make bronze. Bronze was a real game changer, especially when it came to making basic types of farming equipment and tools, or weapons of war.

The next big jump in technology was during the Industrial Revolution (from the mid-1700s to the mid-1800s). That's when *hammer welding* (also known as *forge welding*) was developed. In hammer welding, metal is heated to its plastic state, and then two separate pieces are laid side by side and hammered together. (If you've ever seen a blacksmith at work, you've seen hammer welding in action.)

The next step was based on the discovery of acetylene in the middle of the 19th century. Controlled use of acetylene gas (combined with oxygen) allowed people to cut and melt metals in a way that wasn't possible before. But welding as you know it today came about in the early 20th century, after people had learned how to harness and use electricity. Very basic electric welding equipment and techniques were already being used across the globe at that point, and World War I made it clear that welding technology was going to be critically important for cranking out massive amounts of metal materials, tools, and machinery. Many of the prominent organizations and companies that loom large in the world of welding today got their start during that period. Improvements in welding processes and equipment came in leaps and bounds, and before the first half of the 20th century was over, the world had seen the creation of the major welding techniques that I cover in this book: stick welding, mig welding, tig welding, and oxyacetylene welding, as well as oxyfuel welding and cutting.

Getting Familiar with Metals

Any welding endeavor is much easier if you have a solid working knowledge of metals. The more you know about the metals you're using and how they're likely to respond to the intense heat involved in welding, the more likely you'll be able to manipulate and join them in the way you have in mind for a specific project.

You probably remember from your high-school science class that, like other materials, metals expand when you heat them and contract as they cool off. If you heat them enough, they start to get soft, and eventually (with more heat), they melt. I know that sounds simple, but it's awfully important for welding. Some metals melt at relatively low temperatures, and others have extremely high melting temperatures. A metal's *melting point* is just one of several important properties for welding.

Here are just a few others to consider.

- **Ductility** is a metal's ability to change shape (bend, stretch, and so on) without breaking. Gold has a high level of ductility, while tungsten isn't very ductile at all.
- **Electrical conductivity** is a measure of how well a metal can conduct a current of electricity. Copper conducts electricity really well; by comparison, stainless steel isn't a great conductor of electricity.
- **Strength** is pretty self explanatory: How much external force can a metal withstand without breaking? This one is very important for welding. Steel is a strong metal, but zinc isn't.

You can read up on many more properties of metal, and the more you know, the more easily you can make smart decisions about how to weld those metals effectively.

Not all metals are widely used for welding, of course, and you probably won't work with a huge range of metals in your welding shop until you've been welding for a while. That's completely fine, however, because plenty of exciting welding projects — both fabricating and repairing — involve only a few select metals. (See "If You Can't Beat 'Em, Join 'Em: Understanding Why Welding Matters" earlier in the chapter for more on those divisions.) For example, most of the welding practice exercises I walk you through in this book, as well as the welding projects I detail in Part V, focus on three metals: steel, stainless steel, and aluminum. These three are the most commonly used metals for beginning welders, and you should take the time to get to know them. In the following sections, I give you a quick look at each one.

Steel

Steel is a strong, versatile metal that you'll use all the time in your welding projects. You may not realize it, but steel is really an alloy made up of iron and less than 2 percent of another material. Carbon is often used in steel alloys, and you can find three different levels of carbon steel: low-, medium-, and high-carbon steel. The more carbon in the steel, the stronger the alloy is.

You should use steel in your welding projects when you're looking for a strong metal that's pretty easy to weld and doesn't break the bank when you're buying your materials. You can use any welding process I describe in this book on steel, so versatility is also one of its strong suits. But steel also has its downsides. For one, it's heavy. If you want your fabricated project to be light, steel probably isn't your best bet. Steel is also prone to rusting and *scaling* (flaking off due to oxidation), so you have to spend a fair amount of time cleaning it up (often with a grinder) before and sometimes during welding.

Stainless steel

Stainless steel is amazing stuff. It has a lot of the good qualities that regular steel has (see the preceding section), but it also offers one added bonus: It resists corrosion (rust, for instance) like a champ. You can put a piece of stainless steel out in the yard and let it get rained on for six weeks, and when you bring it back inside it probably won't have a single spot of rust on it. Incredible!

How does stainless steel provide such remarkable resistance to corrosion? Its alloy contains 10 to 30 percent chromium (the rest is iron, although sometimes other metals, such as nickel, are also added to the alloy).

You can weld stainless steel with all three of the major types of arc welding (stick, mig, and tig). It's a great choice if you want your project to resist rusting or to have *hygienic* surfaces (those that don't harbor bacteria and other microscopic critters).

Stainless steel is pretty expensive compared to other commonly welded metals, so be prepared to open your wallet a little wider if you choose stainless steel for a welding project.

Aluminum

Like stainless steel, aluminum is great at resisting corrosion. And aluminum offers another pretty terrific characteristic: It's lightweight. Compared to steel and stainless steel, aluminum is a real featherweight.

Pure aluminum is a popular choice for welders, but aluminum alloys are also frequently used. Copper, manganese, and zinc are just a few of the metals that are often alloyed with aluminum to produce enhanced characteristics in the finished product.

If you're going to be welding aluminum, I recommend going with tig welding. It just makes for a cleaner, easier job. If tig isn't an option, take mig welding; you *can* stick weld aluminum, but it's not ideal — your choices for stick electrodes are going to be limited, and you're probably going to have a difficult time maintaining the correct arc length.

Taking the Time to Understand Welding Safety

Welding utilizes some pretty extreme forces and materials. Most modern welding requires tremendous amounts of electricity, which of course can create a risk for electric shock. No matter what kind of welding you pursue, you're always going to be working around some incredible levels of heat, too, and those kinds of temperatures can harm you, other people, and your property in myriad ways. The metals you weld are sometimes sharp and often heavy, so with them you can get that rare and unfortunate double threat for lacerations and back injuries. Finally, you can't forget other potential hazards that welding can create, including rays that can do serious damage to your eyes and fumes that can hurt your lungs and make you very sick.

Welding is a safe endeavor if you follow all the necessary precautions and respect the equipment, materials, and process. I know as well as anyone that welding involves a lot of potentially hazardous elements, but I also know that if you make maintaining a safe welding environment your first priority, you can weld for years and years without suffering any serious injuries or loss of property. You just have to follow the safety rules and keep your head on straight.

As you work your way through this book I ask only one favor of you: Please read Chapter 3 (on welding safety) carefully and thoroughly. Even if you think you understand welding safety, taking a few minutes to review the key steps for creating a safe welding environment for yourself and others can't hurt.

Exploring Welding Methods

You can use heat to join metals in several different ways, but by far the most common welding methods used today are the arc welding methods. *Arc welding* is really pretty simple in theory: A large amount of electricity creates an arc between an electrode and a base metal, and that arc generates enough heat to melt the materials in the weld area and join them together to make a weld. In practice, however, arc welding includes three different welding processes (stick, mig, and tig) and has many different variables. For example, some kinds of arc welding use a shielding gas, while others don't. The electrodes that you use in arc welding may be *consumable,* meaning they get melted and incorporated into the weld, or they may be non-consumable. The electricity used in arc welding is the source of many other variables, including amperage (which can vary a lot) and current (either alternating current or one of a couple different forms of direct current).

Because the three main types of arc welding are the most commonly used throughout the world and the easiest to pick up, those are the three that I devote the most attention to in the following sections (and throughout the book). However, they aren't the only game in town, so I also include some information on those other types in case you want to branch out a bit.

Stick welding

Stick welding (also called *shielded metal arc welding* or SMAW) is an arc welding technique that has the distinction of being the most commonly used welding practice in the United States today. (More than 40 percent of all welding done now in the United States is stick welding.) The prevalence of stick is even stronger in construction; more than half of all construction-related welding uses stick. And the percentage is even higher in the maintenance industry.

Stick welding enjoys such popularity for three primary reasons. First off, it's cheap. You can get into stick welding for less money than you'd spend to get started with tig welding. Secondly, stick welding is highly portable. The equipment is lightweight, and you can easily use it outdoors if the conditions allow it. Finally, stick welding is versatile. You can use it to work on metals with a wide range of thicknesses, and you can stick weld in just about any position that fits with your skill level.

Stick welding is great, but it isn't perfect. One main reason is that it's messy. Welding waste products, such as *slag* and *spatter,* get thrown around during a stick weld a lot more than they do when you're tig or mig welding. Because of that, you have to plan on spending some time cleaning up your welds and weld area after you're done stick welding. Another of stick's imperfections is its speed (or lack thereof). You have to be pretty good at stick welding to do it quickly (especially compared to, say, mig welding).

You can read all about the stick welding process in Chapters 5 and 6, but generally speaking, stick welding utilizes a consumable electrode with a solid metal rod in its core that melts down and forms part of the weld. Small globules of molten metal flow from the tip of the electrode through the electric arc to the molten weld pool. The electrodes have a coating of *flux* that protects the molten metal from impurities in the air that can contaminate the weld as it cools.

Mig welding

Mig welding is another arc welding technique. You may also hear mig welding referred to as *gas metal arc welding* (GMAW) or *wire welding.* Mig welding is becoming more and more popular, for several reasons. At the top of the list is the fact that most people find mig welding to be easier to pick up than stick and tig. Another big reason is the speed; done correctly, mig welding can be quite a bit faster than stick or tig welding thanks to its continuously fed wire electrode, which doesn't require changing nearly as often as the stick electrodes used in stick welding. You can just keep right on welding without having to stop and change your electrode. Over the course of a welding project, that can definitely save you quite a lot of time.

Proponents of mig welding also cite the low amount of slag and spatter that mig produces. That makes for a more pleasant welding experience, and a much more pleasant cleanup experience. The low chance of *distortion* (unwanted changes in a piece of metal's shape) is also trumpeted by those who love mig welding. Because the process is faster, you don't need to apply as much heat to the weld area for as long, so the metal is less likely to bend and twist in nasty ways.

Of course, mig welding also has its downsides. For starters, mig welding equipment is more complex than stick welding equipment, so it's quite a bit more expensive. The handheld part of the mig welding equipment (called the *mig gun*) is often big and bulky, so it's usually tough to mig weld in tight spaces. Mig welding also relies on the use of a shielding gas to keep atmospheric contaminants away from the weld area, so the process doesn't really work very well outdoors (especially with any kind of breeze).

I save the details of the mig welding process for Chapters 9 and 10, but generally speaking, here's how it works: A wire feeder continuously feeds the wire electrode to the weld area at a speed you control. That produces a steady molten stream that you can easily direct however you want on the surface of the metal you're welding. The weld is completely covered with a shielding gas (usually argon) to prevent impurities from fouling up the quality of the weld; you control the flow of the shielding gas to suit your project's needs.

Tig welding

The last type of arc welding is *tig welding,* which is sometimes called *gas tungsten arc welding* or GTAW. One major advantage to tig welding is that it's extremely clean. If you're tig welding correctly, you may very well go through an entire project without having to spend any substantial amount of time cleaning up. Tig is also extremely versatile. You can use tig welding to work on a lot of exotic metals that just aren't in play for, say, stick welding.

Tig welding has two big drawbacks. One is cost — you can definitely spend a pretty penny on tig welding equipment and supplies, even for start-up. The second drawback is lack of speed. You get a lot of precision out of tig welding, but you pay for it with time.

The tig welding process was originally developed in the 1940s to join aluminum and magnesium, but you can use tig welding to join all kinds of different metals. The big difference in tig welding is that it uses a non-consumable electrode that's almost always made of tungsten. It also requires the use of a water- or air-cooled torch, which holds the tungsten electrode and is connected to the welding machine by a power cable. Like stick welding (see the earlier section), tig uses an arc of electricity to heat metal to its melting point, and you manipulate the puddle to join metals together. The major difference is that tig welding uses a tungsten electrode. You can read more about tig welding in Chapters 7 and 8.

Other welding methods

There's more than one way to skin a cat, and there are more welding processes beyond the big three arc welding techniques (see the preceding sections). Here are a few to consider; check out Chapter 13 for more info.

- **Brazing** is unique among the welding processes because you can use it to join different materials (two different metals, for example). It uses gas rather than electricity, and the heat used in brazing surpasses 800 degrees Fahrenheit.

- **Soldering** is a form of welding that uses (relatively) low amounts of heat. You can solder at temperatures below 800 degrees Fahrenheit. (That's downright chilly when it comes to welding.) You can solder with gas or electricity, but the electricity you use in soldering isn't the same as the type of electricity you use in arc welding. Instead, soldering uses an electric soldering iron that heats up and melts the filler materials you're adding to the project you're working on.
- **Oxyfuel/oxyacetylene welding** is probably the most common gas welding process. You do it with a gas-powered flame that melts the base metal and any filler materials necessary to make the weld. The equipment used for this type of welding is the most portable and low cost in the welding world.

Looking at the Future of Welding

The need for skilled welders is huge right now, and it's only going to continue to grow. New metal alloys are being created and used for a wide range of purposes every day. The industries that rely on welding are expanding rapidly across the globe, and the need for metals to be joined in skillful ways isn't going anywhere in the near future. Welding is a versatile field that you can study in a number of different ways, from on-the-job training to education at a vocational or technical school. If you practice and develop your welding skills and work hard, you can more than likely make a career out of welding. And after you've been a welder for a while, you can very easily transition into a position as a foreman, inspector, or welding supervisor, just to name a handful of the possibilities.

But don't think that you need to make a career out of welding in order to enjoy and appreciate the process. You can weld to fix things around your house, yard, or farm. You can weld to create things that you use in your personal or professional life. You can weld to create works of art or gifts for friends and family. Or you can weld just because it's fun and rewarding (and there are few better reasons to weld than that).

Chapter 2

Considering Commonly Welded Metals

In This Chapter

- Understanding steel
- Getting a grip on stainless steel
- Taking a look at aluminum
- Reflecting on a few other metals

Good chefs know food, good carpenters know wood, and any good welder really knows metal. I know that sounds painfully obvious, but I'm sometimes surprised to learn that many seasoned welders aren't all that familiar with the metals they work with on a daily basis. I think having a nice, rounded understanding of the metals you weld is important, and that's what this chapter is all about.

If I had to pick three metals that most new welders want to begin working on as soon as possible, I'd have to go with steel, stainless steel, and aluminum. If you can familiarize yourself with those three metals and understand their characteristics as they pertain to the various types of welding, you're well on your way to figuring out how to work with three of the most common, versatile metals out there. With that in mind, I start this chapter by devoting individual sections to each of those three metals. After that, I close the chapter with a quick look at some of the other metals that you may want to consider welding, just to clue you in on some basics and give you a feel for what's possible beyond the old standbys.

Steeling Yourself for Using Steel

Steel is an extremely common metal. It's all around you — chances are, you're probably not more than a few feet away from something made of steel. For that reason alone, it's a very important metal for welding.

Steel is an alloy that's made up primarily of iron, along with less than 2 percent of another material. That material is usually carbon, and the amount of carbon present in the steel is an important feature. Here's a quick look at the three different levels of carbon steel.

- **Low carbon or mild steel** has less than .2 percent carbon. This category of steel is extremely easy to work with; you can cut and form low carbon or mild steel a lot easier than many other metals. Lots of objects, including screws, bolts, nuts, and washers, for starters, are made of low carbon steel. Sheets of low carbon or mild steel are often used to make automobile bodies and other familiar products.
- **Medium carbon steel** has .25 percent to .55 percent carbon, and it's more difficult to work with and form than low carbon steel. You can find medium carbon steel in some of the same products made of low carbon steel, but the medium carbon versions are stronger. Machine parts (gears, axles, levers, and so on) are also often made out of medium carbon steel because of its strength and durability.
- **High carbon steel** is the really tough stuff. More precisely, it contains between .55 percent and 2 percent carbon. It's the hardest and strongest type of steel, but it can be a real pain to cut and form. Manufacturers use high carbon steel to make things like cutting tools, files, and hammers because those items need to be strong enough to keep their shapes and integrity through years of heavy abuse.

Getting a handle on forms of steel

Steel is manufactured in many different forms, and each form has its own use for welding projects. Here are a few of the more common forms that you're likely to run into as you weld.

- **Flat steel** is exactly what it sounds like — a flat piece of steel. It's also called *sheet steel.* It comes in a range of thicknesses and sizes, but when it's larger than 12 inches wide, it's called *plate steel.*
- **Steel bars** are made in an array of shapes, but the most common are round, square, or flat. You can see some examples of steel bar shapes in Figure 2-1.

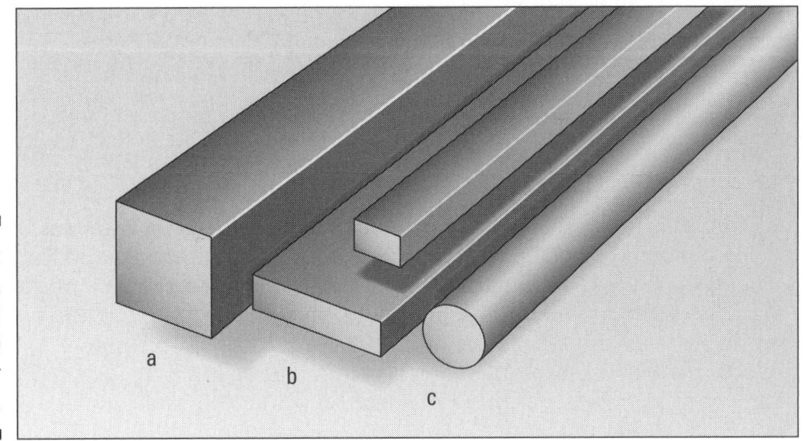

Figure 2-1: Square (a), flat (b), and round (c) steel bar shapes.

- **Rolled steel** comes in two forms.
  - **Hot rolled steel** is made to its finished size while the steel is still red hot. *Iron oxide* forms on the hot steel after it's rolled. It's a grayish-black coating that helps protect the steel from rusting. Hot rolled steel is used for piping, tubing, tanks, and other products.
  - **Cold rolled steel** is made by rolling the steel to its finished size after it's cooled to room temperature. It doesn't get the iron oxide that hot rolled steel gets, so cold rolled steel is smooth and bright looking. It's used for making things like nails and screws.

Like all other metals, steel goes through some changes when you apply the high levels of heat to it that are necessary for welding. The steel around the weld area is subject to distortion and cracking due to the expansion and contraction caused by all that heating and cooling. The good news is that the electrodes and filler metals you use when welding steel are designed to be just as strong (or even stronger) than the metal you're welding, as long as you let the weld cool off on its own after you're done welding.

Knowing when steel is appropriate

Steel is such a ubiquitous metal that it's hard to imagine life without it. It has become an important part of everyone's life, and the ways you live, play, and travel certainly wouldn't be the same if steel was no longer available.

You can use steel for a wide variety of welding projects; it's cheap, readily available, and pretty easy to weld. If you're looking to weld objects such as farm equipment, tools, cars, automotive equipment, specialty containers (drums, pipes, and boilers), or even bridges and parts of buildings, you should strongly consider steel as your metal of choice.

The many properties of metals

It's a little hard to believe how many different kinds of metals are out there. Metals run the gamut from tungsten, which is one of the hardest materials on Earth, to mercury, which is a liquid at room temperature. Here are a few of the physical properties of metals that combine in different ways to make metals unique.

- **Strength:** How much external force the metal can take without breaking.
- **Ductility:** The ability to change shape without breaking.
- **Magnetism**: Some metals (like steel) are magnetic; others (like aluminum) aren't.
- **Hardness:** The resistance of a metal to being damaged when another metal is applied to it.
- **Resistance to oxidization:** When metals combine with oxygen, they become *oxidized.* That's what causes steel to rust, for example. Some metals — tungsten, for instance — are very resistant to oxidization.
- **Electrical conductivity:** Some metals conduct electricity much more efficiently than others. For example, silver is an incredible conductor of electricity, but stainless steel doesn't conduct well at all.
- **Melting point:** Every metal has a *melting point*— the temperature at which the metal turns from a solid to a liquid. This property is critical in welding because, of course, you're trying to melt metal. Tungsten has the highest melting point; you need temperatures of 6,170 degrees Fahrenheit to melt tungsten. That's remarkably high compared to, say, tin, with a melting point of only 450 degrees Fahrenheit.

When is steel not appropriate for welding? Well, if you're working on or repairing a piece of metal that's definitely not made of steel, such as repairing an aluminum piece on a boat, using steel won't work for that project.

If you're starting a new welding project from scratch and really need to end up with a lightweight product, steer away from steel. Steel is durable and cheap, but it's also pretty heavy compared to many other metals. Steel also doesn't work if you're welding something to be used for any sort of food service application; food almost always involves water of some sort, and water rusts steel, creating an unsanitary environment. For the same reasons, using steel for anything in a medical setting is also a bad idea.

For more great information on steel, check out the American Iron and Steel Institute's Web site at `www.steel.org`.

Preparing steel for welding

All metals have to be cleaned and prepared before you weld them, and steel is definitely no exception. Quality welds aren't going to happen if your steel is covered with surface contaminants. And some of those contaminants can be downright dangerous to your health if they're heated up and converted to fumes while you're welding. (Flip to Chapter 3 to read all about the safety gear you can get to protect yourself from fumes.)

You especially need to do your best to remove rust from the surface of your steel. Rust is especially common on mild steel (covered earlier in this chapter), and it can wreak havoc when you're trying to produce a high quality weld. The most common defect you experience as a result of rust is *porosity* (the presence of lots of little holes) in your welds, and porosity can really ruin a good weld joint.

If you want to use a steel that's less prone to rusting, try one that has a little chromium added to it. Chromium slows down corrosion processes in steel. You can also try *weathered* steel, which has a copper alloy in it and holds up well outdoors.

You can use one of two methods for cleaning your steel prior to welding: chemical or mechanical. The method you choose depends on the metal type, the condition of the metal, the welding process you're planning to use, and the equipment available.

Don't assume that a piece of steel (or any metal for that matter) is clean just because it looks clean. Even a new piece of steel fresh from your welding supply store has contaminants on the surface that you need to clean off prior to welding.

Cleaning steel with mechanical methods

When you clean your steel mechanically, you clean the surface by scraping, brushing, or grinding. I usually clean pieces of steel by hand (without power tools) only when the pieces are very small because power tools are just faster otherwise. If you opt to clean steel by hand, I recommend using a very sturdy wire brush. (Check out Chapter 4 for information on wire brushes and other basic tools you need for welding.)

When cleaning a metal before welding, make sure that the metal in your wire brush or the metal on the attachment you plan to use on your power tool is the same as the metal you're going to weld. For example, if you're going to be cleaning a piece of steel, make sure you use a steel wire brush or grinder, not one made of brass or another metal. Otherwise, you run the risk of contaminating the metal with another metal. The exception? It's okay to use a stainless steel wire brush on aluminum.

Tools powered by electricity or pressurized air have become the standard for cleaning steel in many welding shops, especially for shops that weld large pieces. The most common power tools used for cleaning steel are angle grinders, shown in Figure 2-2. An angle grinder in use is depicted in Figure 2-3.

Figure 2-2: A typical angle grinder.

Figure 2-3: Cleaning steel with an angle grinder.

You may think that you can ease up a little on your safety precautions when you're cleaning a piece of metal but not yet welding it, but don't be fooled. You can be injured during the cleaning process, particularly if you're using power tools to clean the metal. Head to Chapter 3 for the lowdown on safety.

Cleaning steel with chemicals

Chemical cleaning is, as you can probably guess, a way to clean steel by using harsh chemicals. I'm talking about some pretty rough chemicals here, so if you go the chemical cleaning route, please be sure to treat the materials with the utmost respect.

The chemicals used to clean steel before welding are caustic and extremely dangerous. They can do serious damage to your eyes and skin. Before using any of these chemicals, be sure you've read and understand the *Material Safety Data Sheets* (MSDS) that lay out their risks. Pay special attention to the ways in which you combine these chemicals because many of them react violently if you mix them up.

You can clean steel with several different chemicals, and here are a few that you may want to consider.

Almost all the solvents used to clean steel and other metals before welding are extremely flammable, so be sure that no spark or flame ever comes in contact with the solvent or its fumes. If you clean steel with one of these solvents, make sure the metal is completely dry and the solvent has evaporated or dissipated before you begin welding. Also, be sure to keep your solvent containers in a safe place away from your welding area, as I note in Chapter 3.

- **Acetone**: This harsh chemical removes rust and oxidation from steel. Welders usually use acetone by spraying it on and rubbing or wiping it off with a clean rag. Compared to some of the other solvents sometimes used to clean metal, acetone is relatively mild, but it's still a toxic material, so proceed with caution.
- **Alcohol:** You can use alcohol for light cleaning and for degreasing. I recommend spraying it on and wiping it off with a clean rag or cloth.
- **Acid**: You can use sulfuric or hydrochloric acids to remove rust scales and oxidation from steel before welding it. These substances are dangerous, so use extreme caution if you go this route — in fact, I recommend cleaning with acid only as a last resort. After applying the acid, make sure you rinse the area thoroughly with hot water and dry it as soon as possible.

Exploring steel welding methods

You can weld steel with just about any of the welding methods I describe in this book, but I recommend sticking with the big three arc welding techniques — stick, mig, and tig. Here's a quick look at a few steel welding specifics to keep in mind for each of those processes.

Stick welding

Stick welding works just fine for joining steel. It's extremely portable — which comes in handy when you need to weld a huge piece of steel that you can't easily move — and as you can read in Chapters 5 and 6, stick welding is versatile and inexpensive compared to the other welding methods. You can use stick welding on virtually any piece of steel that's ⅛ inch thick or thicker.

Mig welding

You can mig weld steel in any position, and that's one of the reasons mig is the most popular choice for steel welding projects. Another reason: You can get a very smooth steel weld with mig, and you can also avoid a lot of the spatter and waste that you may get with stick welding. You'll have a hard time finding a piece of steel you can't weld with mig; very thick pieces of steel work out fine as long you have a powerful mig machine, and you can even weld extremely thin pieces of steel with mig welding.

Tig welding

Like mig welding, tig allows you to join all different thicknesses of steel. Tig also affords the advantage of a very clear view of your welding area, because tig produces very little smoke or *slag* (waste).

If you choose tig for your steel welding jobs, make sure you have your equipment set up correctly. Ask your welding supply shop for the charts that help you determine the correct amperage, shielding gas flow rate, and polarity for tig welding steel. You also need to be sure that you use the right *filler rod* (a metal rod that melts and becomes part of your welding puddle to add strength to your weld), or the quality of your weld may suffer greatly.

Don't use oxyacetylene filler rods when you're tig welding. The copper coating creates impurities in the finished weld.

Going with Stainless Steel

Stainless steel is pretty incredible stuff. It's remarkably durable and strong, and as a result it's used in countless applications all over the world. I doubt you'll be able to practice welding for long before some stainless steel object catches your eye as a possible project component (if it hasn't already).

Understanding the differences between steel and stainless steel

Stainless steel is different from regular steel because it contains 10 to 30 percent chromium. Chromium is added to the steel to create the resistance to corrosion that makes stainless steel so famous (and useful!). Stainless steel can contain a few other substances that make it perform at an even higher level; nickel is the most common.

Stainless steel today is classified into two general series.

- The *200 to 300 series* includes stainless steels made with chromium, nickel, and sometimes manganese. They're more resistant to corrosion than the 400 series, and they generally have better qualities for welding.
- The *400 series* includes stainless steels that don't contain nickel. They can't be hardened as much as the 200 to 300 series.

Stainless steel's most notable (and desirable) quality is its resistance to corrosion. The combination of steel and chromium creates an outer surface that's terrific at resisting rust. Because of that quality alone, stainless steel is used in a huge variety of applications, from beer kegs to hypodermic needles to the St. Louis Gateway Arch.

Deciding when to use stainless steel

Why would you want to use stainless steel in one of your welding projects? Here are a few examples.

- **You need part or all of the project to be rust-proof.** This goal is pretty self-explanatory, but if your project is going to be located outside — especially in an area that gets a lot of precipitation or salt abuse — and you can't let it get rusty, stainless steel is one option to consider.

- **You don't mind spending some extra money.** Compared to other metals (carbon steel, for example), stainless steel is expensive! It can cost as much as five times more than steel.
- **You need part or all of the project to be hygienic.** The strong, consistent surface of stainless steel means that it doesn't have tiny irregularities in the surface that bacteria and other critters can cling to. It also means that you can clean the surface with some pretty powerful cleaners without damaging the material. For those reasons, stainless steel is a popular choice for food preparation and storage equipment, and also for medical purposes.

Looking at stainless steel welding methods

If you want to weld stainless steel, your three best options (by a mile) are stick welding, tig welding, and mig welding. Generally speaking, stainless steels are slightly more difficult to weld than carbon steels. That's because stainless steels have lower melting temperatures, and more *thermal expansion* (they expand more than steel when heat is applied).

- To stick weld stainless steel, you have to use a flux-coated electrode, which protects the metal from the air while you're welding and helps make the weld even more corrosion resistant. Keep in mind that stick welding is the messiest of the arc welding methods, and that can be a drawback when you're welding stainless steel. The random arc marks and spatter caused by stick welding can compromise stainless steel's pristine surface and appearance.

  When selecting stick welding electrodes for stainless steel welding projects, make sure you pick electrodes that have a -15 or -16 suffix. For example, a commonly used electrode for stick welding stainless steel is E-308-16. For more information on picking out stick welding electrodes, take a look at Chapters 5 and 6.
- Tig welding stainless steel is a great option when you're looking to weld thin pieces. I recommend using argon as your shielding gas. You can read more about shielding gases for tig welding in Chapters 7 and 8.
- Mig welding is nice and fast, so it's the ticket if you want to weld thicker pieces of stainless steel. In those cases, tig welding can just take too long.

Working with Aluminum

If you're looking for a strong, lightweight metal, chances are your search can stop at aluminum. It's a remarkably versatile material that's usually at or near the top of the list of most-welded metals. If you stick with welding for very long, you're probably going to want to weld some aluminum, and I can't blame you. Read on to find out more about welding this wonderfully dynamic metal.

Perusing the properties of aluminum

So what makes aluminum so special? For starters, it reacts with oxygen in the air and produces a very tough oxide film on the surface. This aluminum oxide is extremely durable, and it effectively protects the metal underneath it. (That's why aluminum is so resistant to corrosion.)

The aluminum oxide film that covers pieces of aluminum that have been exposed to the air for long periods of time has a higher melting point (3,600 degrees Fahrenheit) than the pure aluminum underneath it (1,200 degrees Fahrenheit). For this reason, you have to remove the film before you can weld aluminum.

Aluminum conducts heat about five times better than steel, and interestingly, it doesn't change color when you heat it up. (Steel, of course, turns all kinds of different colors as you heat it up to different temperatures.) It weighs about one-third as much as steel.

Pure aluminum is a popular choice for welders, but many different aluminum alloys have also been developed. A classification system of four-digit numbers has been created to identify the various aluminum alloys. The first digit is the really critical one because it tells you what is included (other than aluminum, of course) in the alloy. Here are the details.

- **1XXX**: Unalloyed aluminums made up of at least 99 percent pure aluminum
- **2XXX**: Copper
- **3XXX**: Manganese
- **4XXX**: Silicon
- **5XXX**: Magnesium
- **6XXX**: Silicon and magnesium
- **7XXX**: Zinc
- **8XXX**: Other materials (tin is one example)

The two types of aluminum you'll probably weld most often are pure aluminum and aluminum alloy 3003.

Recycle your old aluminum! Almost two-thirds of the aluminum used today has been recycled, and that's the environmentally responsible thing to do.

Eyeing aluminum welding techniques

If you ask me, welding aluminum isn't any more difficult or problematic than welding steel. It can take some getting used to, but after you have the hang of it, you can weld aluminum quickly and efficiently by using any one of the three major arc welding methods: stick, tig, and mig.

- My first choice for welding aluminum is always tig welding. You need to use alternating current (AC) with continuous high frequency to get the best results. (If those terms don't make any sense, flip over to Chapters 7 and 8 for more information on tig welding.) And as with all other tig welding endeavors, make sure you select the right tungsten electrode and shielding gas for the job.
- If you're going to be welding thicker pieces of aluminum, consider going with mig welding. As with stainless steel (see "Going with Stainless Steel" earlier in the chapter), aluminum can take a long time to weld if you're working with thick pieces, and mig welding makes the process go faster. Be sure to keep your electrode wire clean, use a 30-degree leading travel angle, and go with pure argon for your shielding gas. (If those details are Greek to you, check out Chapters 9 and 10.)
- So what about using stick welding for aluminum? Well, it can be done, but I wouldn't recommend it if you can go with tig or mig instead. Not many stick welding electrodes are available for working with aluminum, and maintaining good arc stability is tough. On top of that, you have to fight a constant battle to make completely sure that you keep the covering on the electrode extremely clean and dry, or you're sunk. Bottom line: Avoid stick welding aluminum unless you simply have to do it.

Considering Other Metals

You can do a whole lot of welding with only steel, stainless steel, and aluminum, but you'll inevitably want to branch out a little and work on some less commonly welded metals. Here's a quick rundown of some of those metals, and you can find more information on them sprinkled throughout the other chapters of the book.

- **Copper and copper-based alloys**

  Used for: Plumbing, electrical products, roofing, and as an additive to gold and silver to increase the strength of jewelry.

  Welding process: You can use mig or tig welding for copper and copper-based alloys. Mig welding works in all positions and it's especially useful for thicker pieces of copper. If you use mig for welding copper, be sure you use direct current electrode positive (DC+). Go with tig if you want to weld thinner pieces of copper — I recommend using direct current electrode negative (DC–).

- **Magnesium**

  Used for: An alloy in steel. Makes steel resistant to abrasion, so it's perfect for rock crushers, grinding tools, grinding mills, and so on.

  Welding process: Use tig for welding thin pieces of magnesium-based alloys; use mig if the pieces are thicker.

  Some magnesium-based alloys have high levels of zinc in them. Avoid welding those alloys if at all possible because it's extremely hard to keep them from cracking.

- **Nickel-based alloys**

  Used for: An alloy in stainless steel. Nickel in the 200 and 300 series stainless steels are the most resistant to corrosion.

  Welding process: You can use any of the big three arc welding methods for welding nickel-based alloys. Stick welding gives you a weld that's stronger than the base metal. (That's good!) Tig welding a nickel-based alloy involves using a 2 percent thoriated tungsten electrode and argon as your shielding gas. You can tig weld a nickel-based alloy in any position. (If some of those terms look a little wacky to you, flip over to Chapter 7 for some gory tig welding details.) Finally, if you're going to be mig welding a nickel-based alloy, plan to use a 50/50 mix of argon and helium for your shielding gas. Like tig, you can mig weld these alloys in any position.

Precious metals

You may think it sounds crazy, but you can definitely weld precious metals like gold, silver, and platinum. Gold can be *soldered* (welded at a temperature below 840 degrees Fahrenheit) or *brazed* (welded at a temperature above 840 degrees Fahrenheit), and I recommend doing so with a gas torch — check out Chapter 13 for more on brazing and soldering. The same goes for silver. Platinum is a little different; oxyfuel (see Chapter 11) and tig are your best bets when welding platinum. Oh, and you may want to be sure you have a good lock on your welding shop before you start welding metals that cost thousands of dollars per ounce!

Chapter 3

Setting Your Sights on Welding Safety

In This Chapter

- Making sure you have the right safety gear
- Understanding basic safety rules for welding
- Knowing how to prepare for and handle accidents and injuries

What's the most important aspect of welding? The answer is simple: safety. What good does a beautiful weld and a job well done do you if you've hurt yourself (or others) in the process?

Many dangerous elements make up any welding operation. You use massive amounts of electricity to join metals (which are often sharp and heavy) through melting. You're constantly at risk for electric shock, serious eye injuries, and burns. It's not exactly a pillow fight, is it?

You can't change the basic elements that make welding dangerous, but you can (and should, without fail) take every precaution to make sure your welding projects are as safe as possible. Your approach to welding safety should be complete and relentless; just one careless move or lack of safety preparation can result in serious injury or death.

In this chapter, I cover all the aspects of welding safety that you need to understand and remember when you get started as a welder. As I mention many times throughout this book, this chapter is really the most important one, and I hope you read it carefully and take its information to heart.

Welding safety is no accident! Be sure to take the precautions necessary to ensure your safety and the safety of those around you. You also need to make sure anyone who comes near your welding projects is aware of proper welding safety practices so that they can help keep themselves safe as well.

Gearing Up to Protect Yourself

A big part of welding safety is making sure you have the right safety equipment on hand for every job, and that's what I discuss in the following sections. Figure 3-1 shows a welder in full protective gear; make sure you have all these items available before you get going on any welding project.

Figure 3-1: A welder wearing recommended safety gear, including ample eye protection.

Choosing eye protection

To protect your eyes from flying debris (usually metal), which you encounter plenty of if you weld regularly, make sure you wear safety glasses in your shop at all times. Make a habit of putting them on as soon as you walk in the door — that way, you're always protected.

To fully protect your eyes from the damaging effects of ultraviolet welding rays, you need a welding helmet. These helmets protect your eyes (and the rest of your face) from damaging rays, and they include special lenses that allow you to see your work clearly without suffering eye injuries, which I discuss in "Shielding yourself from burns" later in the chapter. (Check out Figure 3-1 for a look at what a welder with a welding helmet looks like.)

You need to get a helmet appropriate for your particular welding work. New welders often buy helmets that don't cover all their needs, or purchase ones that are far too heavy duty (and therefore expensive) for the projects they want to do. The following list explains the features of welding helmets that can help you figure out which helmet is right for you:

- **Shade number:** The lenses in welding helmets are rated according to *shade number.* Shade numbers for welding range from #8 to #14. The higher the shade number, the higher the *amperage* (the amount of electrical energy flowing through a circuit) you can use when welding without damaging your eyes. Most welding operations are carried out at shade #10, but if you get above 140 amps you must get a darker lens with a shade in the #12 to #14 range.
- **Auto-darkening or passive lens:** Many welding helmets now offer an *auto-darkening lens* that automatically increases the shade number to account for more-intense welding rays. The alternative is a *passive lens,* which is the older style of helmet that has a fixed shade number. Auto-darkening helmets are nice because you can see through the lens when you don't have a live welding arc, so you don't have to constantly take your helmet off and put it back on when you're working on a project. However, auto-darkening helmets are much more expensive. Passive lens helmets are cheaper, but you'll likely have to remove them and put them back on over and over while you're welding.
- **Comfort:** Make sure you get a helmet with an adjustable headband that feels comfortable on your head and neck. The helmet should stay in place and shouldn't pinch your head.

Welding helmets take a lot of abuse, but you don't have to replace your helmet if it gets nasty or dinged up. You can always repair or replace parts of the helmet (lenses, shields, and headbands for example) instead of emptying your wallet on a new one.

Sunglasses don't protect the eyes from welding rays. Don't even think about using them as eye protection.

Make sure you do whatever you can to protect the eyes of others who may be near your welding projects. Ultraviolet welding rays are so powerful that they can penetrate through closed eyelids. If you have bystanders, make sure you say "Cover!" loudly before you strike an arc and get started welding. (You need to let your audience know beforehand what to expect when you say "Cover!", of course.)

Keeping the right fire extinguisher on hand

With all the extreme heat and sparks created during a welding job, you shouldn't be surprised if something catches fire at some point. Because fires are a very real threat, keep a fire extinguisher on hand at all times.

Four main types of fire extinguishers (shown in Figure 3-2) work on four different kinds of fires:

- **Class A:** *Class A* extinguishers take care of any fires that produce ash. These fires usually involve wood, cloth, or paper.
- **Class B:** *Class B* fire extinguishers are for fighting fires caused by flammable liquids, such as gasoline, oils, and paints.

- **Class C:** Electrical fires call for *Class C* fire extinguishers. Don't use any water- or foam-based extinguishers on an electric fire, or you put yourself at risk of serious shock.
- **Class D:** Use *Class D* extinguishers on metal fires. Combustible metals like magnesium, potassium, and sodium are extremely flammable, and fires based on these materials are very dangerous. Class D extinguishers are best for these fires because the other classes often disperse a metal fire instead of putting it out, and that makes the situation worse.

You should keep a Class C extinguisher nearby when you're welding, in case an electrical fire breaks out. I recommend also keeping a Class B on hand for oil fires.

In addition to fire extinguishers, keep a water source (a hose, for example) and a bucket of sand nearby when you're welding. They can come in handy when you need to control ash and metal fires, respectively, without the need for special extinguishers.

The best way to avoid needing an extinguisher is to cut down on the flammability of your project in the first place. For example, you may find the need to weld a metal 55-gallon drum. (It's a common welding task.) If so, take care to ensure that the drum doesn't have residual flammable materials on the interior. Before welding, be sure to clean the interior with steam or a chemical cleaner.

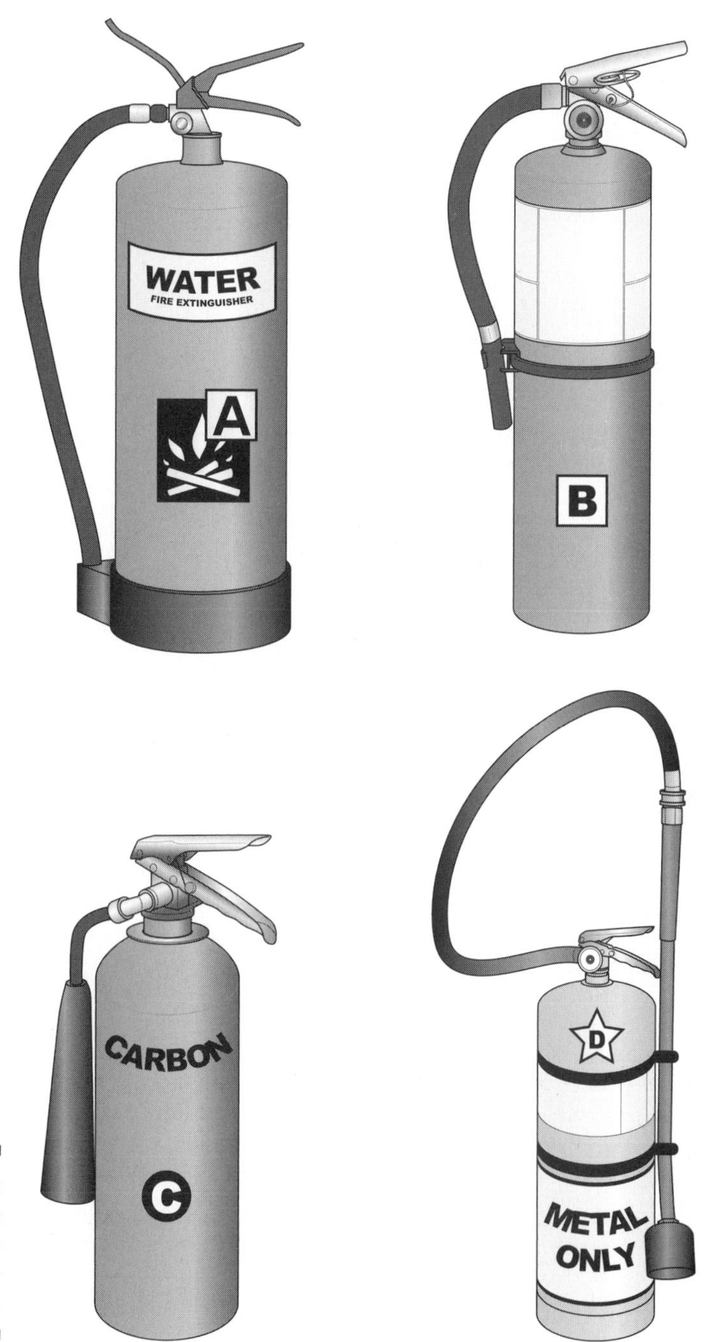

Figure 3-2: Four classes of fire extinguishers.

Wearing protective clothing

If you've ever thought of yourself getting started on your first welding project wearing a pair of shorts, an old t-shirt, and a pair of flip-flops, think again. If you want to keep your body safe from the hazards of welding — especially burns — you have to always wear appropriate protective clothing.

Welding produces a huge amount of hot metal sparks and flying slag (molten metal). To keep your skin from being burned, choose clothing made from the right material. One of the best options is 100-percent wool, but it can be difficult to find and is extremely hot. (Some people are also allergic to wool.) I prefer 100-percent cotton, and it's the most popular material used for welding clothing. The fabric should be thick (because welding rays can penetrate thin fabric and burn your skin like a sunburn) and dark-colored; black and dark blue are good options.

Sorry, but you can't use that old polyester leisure suit from the '70s when you're welding. Synthetic fabrics like polyester, nylon, and rayon are unsafe for welding because they can melt and adhere to your skin when they get hot. And that definitely does *not* promote leisure.

Before welding, check to see that your clothes are free from oil and anything else flammable. They should also be tight fitting and free from holes and frayed or torn edges. Here are a few more clothing-related details to bear in mind.

- **Shirts:** Wear long-sleeved shirts only, and tuck your shirttail into your pants. If you wear a shirt with front pockets, make sure they have flaps covering them so that sparks can't fall down inside. Keep your shirt sleeves rolled down at all times.
- **Pants:** Make sure your pants cover the tops of your shoes or boots so that sparks can't get in.
- **Gloves:** Protect your hands with leather gauntlet gloves. Don't wear cotton gloves at any time.
- **Boots or shoes:** Go with leather boots or shoes, and make sure they come up high on your ankle. Leather is the best option because it protects you from the hazards and heat created by the welding process. The tops of the shoes should be smooth so sparks don't get trapped.
- **Hats or beanies:** You can buy a welder's beanie to keep your hair from catching fire — not a bad idea.
- **Leather jackets:** A full leather jacket protects you from sparks and slag when you're doing work in unusual positions, but many such jackets are very hot and heavy to wear. If a leather jacket is too hot and uncomfortable, you can buy leather sleeves and bibs only. These items are cooler and more comfortable, but they do leave your back unprotected.

- **Leather aprons:** These items protect your lap, and I highly recommend wearing one if you plan to sit down while welding.
- **Spats:** If you're going to have a large amount of sparks and *slag* (the waste products that come from the flux) falling at your feet, you may consider getting some *spats* (shoe covers). Leather spats greatly reduce the risk of burns on your feet and damage to your shoes or boots.

Be careful what you have in your pockets when you weld. If you have a cigarette lighter or matches in your pocket during a welding project and a spark gets close enough, it can cause a fire in your pocket.

Watching health hazards: Using a respirator and Material Safety Data Sheets

Keep the *Material Safety Data Sheets* (MSDS) for every chemical and other potentially hazardous material in your welding shop on hand. These sheets describe all the properties of a given substance and include details on potential health hazards. Make certain that you and anyone who enters your welding area know the dangers of the materials on hand.

If you're going to be working with metals or other materials that can produce toxic fumes — check your MSDS to be sure — you should wear a respirator. Simply welding or cutting certain materials can cause a welding *plume,* which is a mixture of dangerous gases, fumes, and smoke, and that's definitely not the kind of thing you want to inhale if you're interested in avoiding respiratory problems.

Figure 3-3 shows four types of respirators:

- On the top left, you can see a *welder's mask,* which works to keep fumes out of your lungs.
- The mask on the top right has canisters that filter fumes and other hazardous materials while you're welding. You can get different types of canisters to match the kinds of hazards specific to various welding projects.
- The basic dust mask on the bottom left offers the least amount of protection.
- The mask on the bottom right filters out particulate matter, so it's a step up from the basic dust mask but doesn't do much to protect you from harmful chemical fumes.

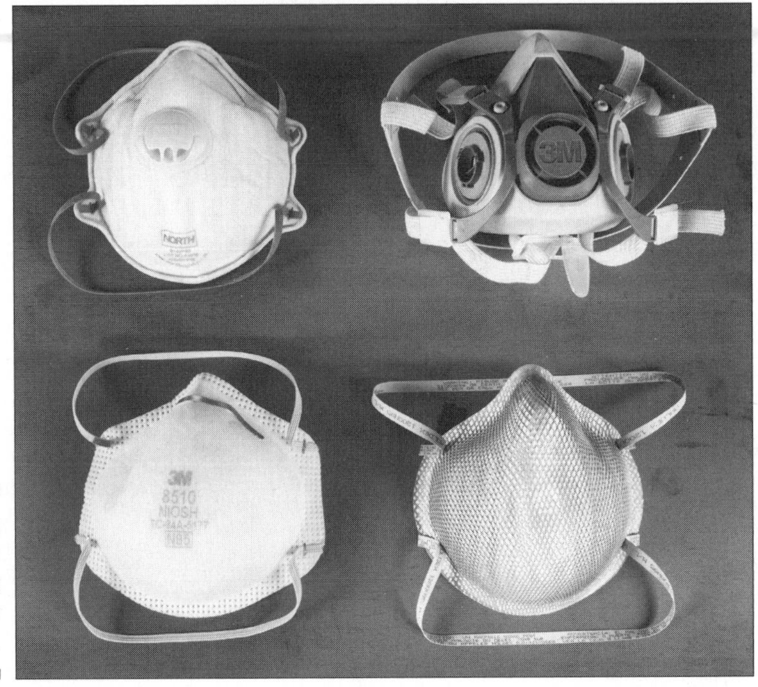

Figure 3-3: Four different respirator masks.

The metals you weld aren't the only items that can cause harmful fumes — the substances that may coat those metals can also be hazardous. If you suspect that a piece of metal may be coated with paint or another kind of chemical that may generate dangerous fumes when heated up, be sure you remove that substance (with a chemical cleaner or grinder) before you weld.

In order for a respirator to be completely effective, you need to fit it with the right filter. Check your MSDS to find out which filter you need to protect your respiratory system from a given substance. You can also ask for help at your welding supply store.

Observing Basic Safety Rules

I've heard that some people are accident prone, but I just don't believe that — especially when it comes to welding. Welding accidents occur when welders act carelessly, don't read instructions, and don't follow safety regulations. Every year, thousands of people are injured because they don't take safety precautions while welding.

In many ways, welding safety is a habit. It's the result of taking the appropriate precautions over and over, every time you weld. Following are a few safety tips that you should make a habit of in your welding shop:

- Don't take chances while welding.
- Don't use any welding equipment or supplies until you've thoroughly read the instructions and any additional safety information.
- Any time an accident occurs, be sure to fix the problem that caused it before doing any additional welding.

In addition to these very basic safety tips, you also need to be familiar with a few more-involved safety goals, and that's what I cover throughout the rest of this section.

Keeping your work space clean

When it comes to your welding area, practice good housekeeping. Take the time before, during, and after a welding project to make sure the area is tidy and organized. Here are a few pointers to help you accomplish the goal of maintaining a shipshape welding shop.

- Designate a place for all tools, and when you're done using a tool, be sure it goes back in the correct place. At the end of a welding project (or at the end of the day), clean all tools and inspect them for any damage that may have occurred while you were welding.
- Keep cables neat and free of knots. Nothing makes for a worse tripping hazard than messy tangles of cables.
- If you have any combustible items in your shop, be sure they're stored away safely before you start to weld. ("Storing flammable liquids and gases" later in the chapter gives you the lowdown on proper safekeeping of such items.)
- Be sure your work area has adequate lighting. A well-lit workspace is less likely to become cluttered and disorganized.
- Keep all scrap metal out of your welding area.
- If a spill occurs, stop what you're doing and clean it up immediately.
- When your waste or garbage container becomes full, take the time to go and dump it.

Checking for leaks

Your welding shop is bound to contain pressurized containers full of liquids or gases. Take special care to ensure that none of those containers (or the tools that you attach to them) has leaks. Checking for leaks is a simple process:

1. **Mix one teaspoon of dish soap in a spray bottle full of water.**
2. **Spray the mixture onto any area of a pressurized container that is likely to have a leak.**

   Fittings and valves are usually the most likely sources of a leak.
3. **Look and listen closely for bubbles.**

   If you see bubbles or hear a bubbling noise, that probably means you have a leak. If the bubbles are popping up near a fitting, use a wrench to tighten up the fitting. Then reapply the soapy water and check again for bubbles. If you see bubbles on or near a valve, it may be time to replace the valve.

If you're using any oxyacetylene equipment (see Chapters 11, 12, and 13), you know when you have a leak if you smell garlic. Manufacturers now add a compound to the acetylene that makes it smell like garlic; if you get a waft of garlic smell and think you may have a leak, use the soapy water method I describe in this section to find out where the leak is coming from. (Of course, the garlic smell may also be the result of the pizza you had for lunch, in which case you may just want to grab a stick of gum.)

White thread seal tape can be useful for making sure that fittings and valves don't leak, but be sure you never use the tape on brass fittings because brass fittings seal themselves.

Hoses are also common sources of leaks, so make sure to inspect your hoses frequently for nicks and burned sections. And please don't try to fix your hoses with tape! Tape fixes aren't reliable, and they don't last very long. Figure 3-4 shows you what a leaking oxygen hose looks like.

Getting the ventilation right

Welding produces byproducts, and while you're welding you can be exposed to a number of dangerous gases. These gases are created when metals are heated above their melting points, when certain electrodes (with special coatings) are in use, and when you don't scrape or clean certain materials (paint, for example) off a section of metal before you weld it.

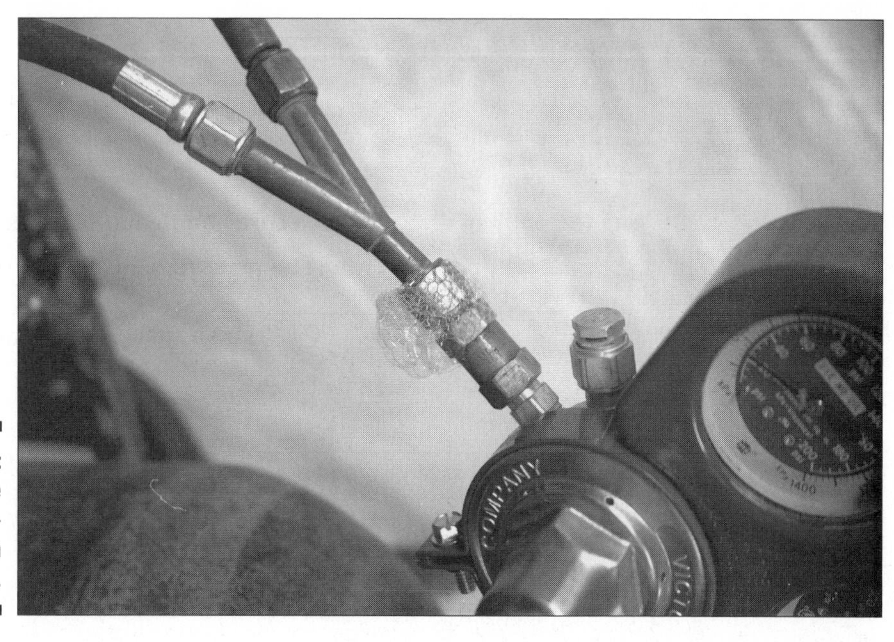

Figure 3-4: An example of a leaking oxygen hose.

If you want to stay healthy, you certainly don't want to inhale those gases, so make sure your welding area is properly ventilated before you begin any welding operation. You can use natural ventilation if you have windows, doors, or garage doors in your welding shop that you can open safely to allow a breeze in. You can also utilize forced air movement with fans and blowers. For detailed information on how to ensure a suitable level of ventilation in your welding workspace, check out the welding-specific information provided on the Occupational Safety and Health Administration (OSHA) Web site at `www.osha.gov/SLTC/weldingcuttingbrazing/`.

Proper ventilation is always important, but it's completely critical when you're welding metals that create particularly noxious, dangerous fumes. These metals include (but aren't limited to) lead, zinc, and cadmium. Be sure to read your MSDS before using any new material.

Storing flammable liquids and gases

You may need to keep several flammable liquids and gases on hand in your welding shop, including acetylene, propane, and natural gas. You need to treat these materials with a great deal of care and respect because they can cause a huge amount of injury and damage in no time at all.

Store all flammable liquids in sealed containers and keep the containers in a flammable liquids locker like the one depicted in Figure 3-5. These lockers are made of metal and have a door that closes securely. You can buy a flammable liquids locker at your welding supply store or even a home improvement warehouse. As you can imagine, you don't want to put the locker anywhere in your shop or welding area with a source of heat nearby.

Figure 3-5: A common flammable liquids storage locker, appropriately labeled.

A few common flammable welding gases (such as the acetylene, propane, and natural gas I mention earlier in the section) are compressed so that a large amount of gas can be put into a cylinder for controlled use. If you have these cylinders in your welding shop, be sure to store them far away from any heat source. Oxygen gas is also very dangerous when pressurized, so treat it with the same respect you would any other potentially hazardous gas.

Knowing your surroundings

You can prevent many different types of safety hazards in your welding shop simply by being aware of your surroundings and reacting quickly when something goes awry. When you're working on a weld, you can easily get really focused and lose track of what's going on around you, so you have to make a real effort to stay aware of your environment.

Pay close attention to any unusual smells or sounds. If you hear or smell something strange, check it out immediately — it may very well be a fire or something starting to catch fire. When you take a break from welding, either for a few minutes or for the night, take some time to look around the area to make sure nothing is burning or smoldering.

When you finish welding, double-check to make sure your welding equipment is turned off so that it doesn't start a fire after you leave it unattended.

You should also take care to understand the materials you work with and give them the respect they deserve. Welding requires a number of harmful materials, and the short- and long-term effects of those items on your health can be severe. As I note throughout the chapter, always look at the MSDS for your materials so that you're well informed of potential risks.

Don't weld or cut anywhere near an area that may contain explosive or flammable vapors.

Welding also creates quite a bit of waste material that can harm your health. *Electrode stubs* (the 2- to 3-inch pieces of stick welding electrode that you can't use) and scrap metal are two of the top offenders. Keep those things in safe containers until you can take them to your local metal recycling center. Some of the flux and dust that you generate when welding can be considered hazardous waste; read the literature that comes with your electrodes to understand what you're up against. If you're dealing with hazardous waste, don't just throw it in the trash, on the ground, or in the toilet. Instead, dispose of it by taking it to your area's hazardous waste disposal facility.

Protecting yourself from electric shock

Most welding equipment uses some sort of electrical power, so electric shock is a very real risk in the welding trade. Electric shocks can cause injury, death, fires and explosions; here are a few tips to keep in mind in an effort to keep your risk for electric shock at an absolute minimum.

- Don't allow electrode holders to touch wet gloves or wet skin.
- When using extension cords, be sure to plug the power tool you're using into the extension cord before plugging the extension cord into the wall outlet. Then be sure you unplug the extension cord before you unplug the power tool.

- Be sure your extension cords don't have kinks, knots, or nicks. Inspect your cords before you use them (every time), and get rid of the cord immediately if you notice damage.
- If a cable or cord feels hot to the touch when you're using it, discontinue use right away. The heat means that the cord isn't big or heavy duty enough for the amount of electricity you're using, and that can be very dangerous.
- If any of the power tools or extension cords you're using has a three-pronged plug and is missing one of the prongs, don't use the tool or cord under any circumstances.
- Keep all power tools and cords dry at all times.
- If your power tools have worn or broken parts, repair or replace the broken parts immediately.
- If your power tools have keys, chucks, or wrenches attached to them, be sure to remove those items before plugging in or using those tools. They can create a hazard if they're still inserted when the power tool starts moving or spinning.
- If a power tool's housing is cracked, don't use it.
- Make sure you provide a ground for any electric tool or device that requires grounding. (Read tool and equipment instructions thoroughly to determine whether you need to provide a ground.)

Shielding yourself from burns

You can suffer burns from two different sources while welding: hot metal and ultraviolet light. Burns caused by hot metal are self explanatory, but you may not realize at first that you can also receive a burn from the ultraviolet light that's a part of arc welding rays.

These rays are the same as what's generated by the sun, except welding rays are much more concentrated. I'm sure as a child you heard to never look directly at the sun, right? Well, you certainly don't want to look directly at a welding arc either. If you do, you can easily suffer burns on your retina, and you may lose part of your eyesight. These types of burns are called *flash burns* or *arc eye,* and as you can imagine, they're extremely painful. You feel like you have something in your eye, and usually that's caused by a blister on your eyeball (fun!). And just because you don't feel the effects immediately doesn't mean you haven't been affected — they start about 8 to 24 hours after you look at a welding arc. Damage can occur even when the light is reflected off of a shiny surface.

To reduce the risk of burns while welding, minimize the amount of ultraviolet light reflected from the welding arc by painting the surfaces of your welding workspace (even the walls) with a flat, dark-colored paint (black and dark blue are good choices). You should also always wear a welding helmet and protective clothing, as I describe in "Gearing Up to Protect Yourself" earlier in the chapter.

Metal you've just welded will be hot, of course, so you need to make sure no one burns themselves on it. If you have to walk away from the weld, even for a minute, write "Hot" on the piece with a soapstone so passerby know not to touch.

Handle all hot metal with pliers to prevent burning your hands, even when wearing gloves. Another habit to develop is using the back of your hand (cautiously) to check to see whether something is hot. Put the back of your hand about a foot away from the potentially hot piece and move it slowly toward the piece. If it's too hot, you'll be able to tell when the back of your hand gets within two or three inches.

Maintaining your equipment

If you want to create the safest possible working environment for your welding jobs, you have to maintain your equipment extremely well. If your welding equipment and tools are falling apart, they're just waiting to cause an accident, and almost all those types of accidents are preventable if you take care of your gear. Here's how you can do just that.

- **Perform a regular check of all equipment.** Keep a list of all your equipment on a notebook or clipboard and do a monthly check through all of it to make sure everything is in good working order. Pay special attention to your electric equipment, hoses, cords, and regulators to make sure you don't get shocked because of faulty equipment, and also to see that you don't have a potentially harmful gas leak.
- **Don't hesitate to get professional repair help.** If you notice something odd or potentially dangerous with some of your equipment and you don't feel confident in your ability to remedy the problem on your own, take the equipment to a qualified service professional. Your local welding supply store should have someone on staff who can make those kinds of repairs, or be able to point you toward someone who can.
- **If a tool or piece of equipment isn't working the way it should, stop using it!** Don't fool yourself into thinking that you can "work out a bug" by just using a tool a little more.

Make sure the heads of your chisels and punches are ground round to keep the chips around the head from flying off when you're using them. That's a common but easily fixable problem.

Being Prepared for Injuries and Accidents

If you follow the safety guidelines I describe throughout this chapter, you greatly reduce the chances of an accident occurring in your welding shop. However, if an accident does happen, you need to know exactly what to do and how to handle it. That's what I tackle in this section.

You should always have a telephone handy in case of an emergency, so you can quickly dial 911. It's also a good idea to let friends and family know when you're working in the shop so that they'll notice if you don't come back within a reasonable amount of time.

Equipping your first-aid kit

Keeping a first-aid kit in your welding shop is always a good idea so that you have the items you need if an accident occurs. Here's what I recommend you include in that kit.

- Sterile bandages of various sizes
- Cold pack
- Sterile gauze in various sizes and lengths
- Burn gel or spray
- Antibiotic ointment
- Hydrogen peroxide
- Sterile compress dressings of various sizes
- Sterile gloves
- Tweezers
- Sterile adhesive tape
- Eyewash
- Antibacterial moist towelettes
- A protective airway mask

You should be able to find a first-aid kit that includes all these items at a hardware store or home improvement warehouse. I provide a useful example in Figure 3-6.

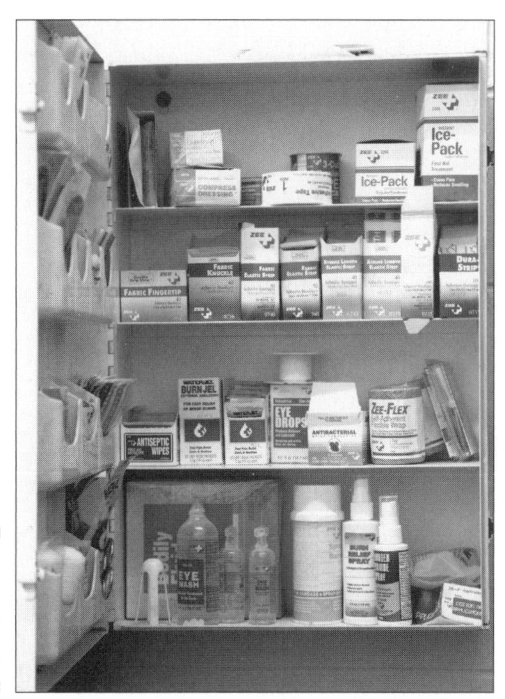

Figure 3-6: A well-stocked first-aid kit.

Knowing how to handle injuries

You may not be a doctor or an EMT, but if an injury occurs, you can still take some very basic steps to begin treatment and keep the injury from becoming more serious. Below I give you an overview of some of those steps.

- **Bleeding and wounds:** Apply steady pressure with a clean cloth or gloved hand. If the wound isn't severe, clean it thoroughly and dress it with sterile bandages. If the wound or bleeding is severe, call 911 or go to the hospital immediately.
- **Broken bones:** Don't move the injured body part or try to reset the bone. If any section of bone is protruding from the skin, cover it with a sterile, moist cloth. Call 911 or go to the hospital immediately.

- **Electric shock:** If you witness someone suffering from an electric shock, don't touch them! You'll become part of the circuit and be shocked, too. Instead, quickly turn off the source of the power. If the victim is unconscious, not breathing, or otherwise seriously injured, call 911 immediately.
- **Gas inhalation:** If you or someone in your work area inhales dangerous gases, get them into the fresh air and seek medical assistance immediately.
- **Poison swallowed:** If someone accidentally swallows a poisonous substance from your welding shop, call your poison control center or 911 immediately and follow its instructions. Take care to note the details of the poison that has been consumed. Don't let the victim drink anything unless you're directed to do so by emergency personnel or someone from the poison control center.
- **Heart attack:** Welding can be strenuous, and heart attacks while welding aren't unheard of. If you or someone near you begins to experience severe pain in the chest, pain in the left shoulder, aching in the left arm, shortness of breath, or a bluish color in the lips or fingernails, dial 911 immediately.
- **Clothing fires:** Welding causes a lot of sparks, and those sparks can ignite clothing. If your clothing catches fire, strip it off immediately if possible. If you can't strip it off quickly, you should stop, drop, and roll to put out the fire. Don't run around — that fans the flames and makes the fire larger.

  You can purchase a fire blanket to help extinguish small clothing fires or other fires in your shop. These blankets are usually made of fiberglass or wool treated with a fire retardant chemical, and you can buy them at a welding supply store or online.

- **Burns (heat):** Each one of the three different types of heat burns has its own treatment. (You can see examples of all three kinds of burns in Figure 3-7.)
  - *First-degree burns* cause the skin to become red and tender but don't break the surface of the skin. If you suffer a first-degree burn, run cold water over the area to cool it off and then cover it with a clean cloth and apply sterile bandages. Don't use ice on a burn because it can damage the tissue.
  - *Second-degree burns* occur when the skin is severely damaged; they commonly involve blisters and even breaks in the skin. If you suffer a second-degree burn, run cold water over it to decrease the pain and then cover the affected area with a sterile bandage. If a blister forms, don't try to break it.
  - *Third-degree burns* are the most serious type. If you suffer a third-degree burn, you likely feel very little pain in the affected area because the nerve endings have been destroyed. If you suffer a third-degree burn, seek medical attention immediately. If clothing is embedded in the burn, don't try to remove it yourself.

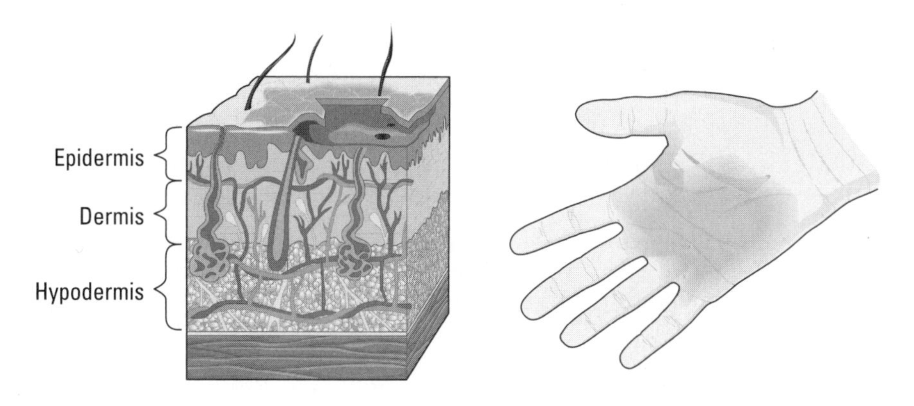

First-Degree Burn

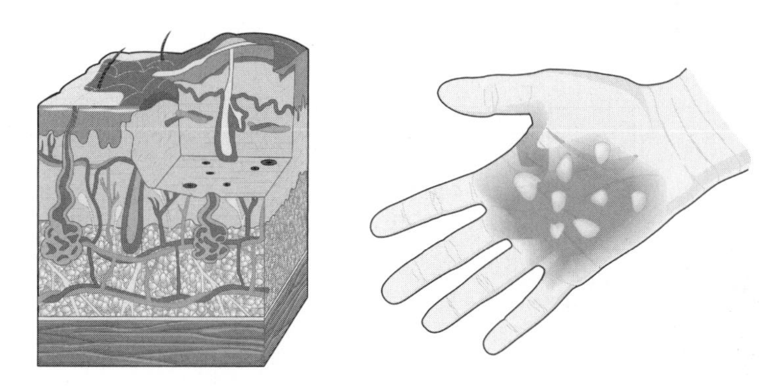

Second-Degree Burn

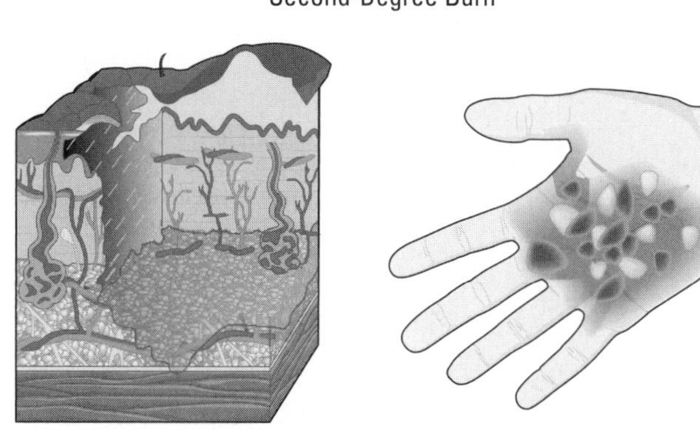

Third-Degree Burn

Figure 3-7: The three different kinds of heat burns.

- **Burns (chemical):** Some chemicals can cause burns; if your skin comes in contact with one of these materials, immediately flush the area with water. Remove all contaminated clothing and cover the burn with a clean, dry cloth. If the burn is serious, seek medical attention.
- **Eye injuries:** If a foreign object strikes or becomes embedded in your eye, place gauze over the eye immediately and secure it with a bandage. Don't attempt to remove the object. Go to an eye doctor, call 911, or visit an emergency room as soon as possible. If a chemical comes in contact with your eye, flush out the affected area with warm running water right away and then place gauze over the injured eye and seek medical attention.

  If you think you may have a flash burn on your eye, see a doctor immediately. You shouldn't try any home remedies, but you can put a cold, wet washcloth over your closed eye to help ease some of the pain. (Flip to "Shielding yourself from burns" earlier in the chapter for more on flash burns.)
- **Plume poisoning:** If your welding area isn't well ventilated and you're not wearing a respirator, you can suffer from *plume poisoning,* which is sickness or injury that results from inhaling dangerous welding byproducts. If you accidentally inhale those substances, you should immediately get away from the plume and into some fresh air. Then seek medical assistance right away. For more on respirators, check out "Watching health hazards: Using a respirator and Material Safety Data Sheets" earlier in the chapter.

Chapter 4

Setting Up Your Welding Shop

In This Chapter

- Selecting and setting up your welding location
- Making sure you have all the necessary equipment

If you want to succeed with any welding endeavor, you need to cover two bases: location and equipment. You can't get very far with welding — particularly *good* welding — if you don't have the right kind of setup, from the building itself to the tools and other gear you need to equip the space properly.

Some folks who are new to welding think they can simply clear out a space in their garages, buy a welder from an online retailer at a deep discount, and start making high-quality welds in no time. Not so fast! You need to think through all the requirements necessary for a welding space that allows you to do good work and do it safely. And when you're talking about welding equipment and tools, you need to make sure you have all the basics (and maybe a few extras as well).

In this chapter, I tell you how to get set up for welding. I start by filling you in on the details of how you can identify a suitable location for welding. No two welding methods or jobs are the same, but even with all that variety, all good welding sites share a number of characteristics that you need to maintain if you want to ensure a productive and safe working environment. After that, I move on to a discussion of welding equipment, which you need to choose carefully so you end up with gear that fits your welding needs.

Choosing a Location

You can't find too many hobbies or jobs that are more versatile than welding. Welders with lots of experience are able to make welds in some pretty amazing places (underwater and in space, for example). But if you're just starting to weld, you can get set up in a relatively simple, straightforward way. You don't have to have a massive, state-of-the-art facility for your welding endeavor; you really just need to make sure your welding space has a few basic features, including the following:

- **Adequate lighting:** Like all other trades, welding requires ample lighting so that you can clearly see what you're working on. Make sure your welding space has plenty of light, whether that's natural, electric, or some combination of the two.
- **Flame retardant floors and walls:** Many welding tasks create sparks, so you certainly can't fire up the old welding machine in a room with shag carpeting and wallpaper. Instead, try to find a space that has, for example, concrete floors and cinder block walls (or something equally resistant to catching fire).
- **Large doors:** If you're going to be welding any large items, you need to be able to get them into and out of the welding area safely and smoothly. Keep in mind that you may limit yourself in terms of the range of welding projects you can tackle if you can't fit certain items into the space. A large overhead or roll-up door is necessary if you want to work on bigger projects indoors.
- **Sufficient electrical service:** Welding machines and other equipment used in cutting, grinding, and welding can use quite a lot of electricity, so be sure that your welding location is properly serviced. Generally speaking, you should be in good shape if you have 220 single phase service (the kind of electrical outlets you use to plug in an electric clothes dryer) or 110 single phase service (the kind of outlets located all throughout your home, used for plugging in lamps and small electrical appliances) with 20-amp circuit breakers. (Those are some of the most common circuit breakers out there.)
- **Ventilation:** You can read more about safety-related issues like ventilation in Chapter 3, but for this part of the discussion, just make sure your welding space is properly ventilated so that you (and any other people nearby) don't suffer any negative health effects from the gases and smoke that welding can produce.

Deciding how much space you need

The amount of space you need in order to weld safely and successfully is always important. Because of the potentially hazardous nature of many welding processes, you need to make sure you have enough space; otherwise, you put your health and the quality of your work at risk.

Many people who are new to welding assume that they can set up a little welding shop somewhere in their home. (A garage or shed is usually the first choice.) That can be okay, but only if the space you're considering fits the criteria in the preceding section and also doesn't compromise your safety. (Be sure to read Chapter 3 carefully if you have any doubts or questions about the safety precautions you need to take before you begin welding.)

Be sure that the space you're welding in doesn't contain any flammable material located within 35 feet of the actual welding area. That may sound like a large distance, but you may be very surprised (in an unpleasant way) how far sparks can fly when you're making some welds.

When you're considering a potential space for your welding work, think hard about the kinds of items you plan to work with. Is your space big enough to allow you to move those objects around without banging into other projects, your tools, or — worse yet — your body? Keep in mind that the equipment you're working with may very well be heavy, sharp, and/or hot. Those aren't the kinds of items you want to be trying to manipulate and move in a cramped environment.

Another important factor to consider is the amount of space your welding equipment takes up. Most welding machines — more about those later in this chapter — are only about three feet by four feet, but you may end up eating up quite a bit of space after you factor in all the other gear and supplies you need. Be sure to allow plenty of room for your equipment, or your welding projects may suffer from a cumbersome and disorganized workspace.

Contemplating indoor versus outdoor

Sometimes the welding project you're working on dictates the location in which you're forced to work. If you're welding a big piece of equipment that has broken and become immobilized outdoors somewhere, you may very well be forced to do your welding outside. If that situation (or something similar) arises, don't automatically think that it presents a difficult challenge. Welding outdoors can actually be just as enjoyable and productive as welding in an indoor welding shop, and in some cases even more so. Indoor and outdoor welding each have a number of advantages and disadvantages you want to bear in mind when you find yourself in a situation that allows you to make that choice on a particular welding project.

When you weld indoors, you often have the advantage of regulating the temperature in the building. That can be a huge plus, as anyone who has been forced to weld outside in the sweltering heat of summer can attest.

You can also control the lighting to fit your needs and the needs of the job, and when you weld indoors, your equipment isn't exposed to the damaging effects of the weather.

However, welding indoors isn't all positive. No matter how big your welding space is, some jobs just don't fit indoors. Welding inside a building can also be more difficult if you're doing the kind of welding that produces an unusually large amount of smoke and dangerous fumes. Your indoor welding space has ventilation, of course, but chances are your indoor ventilation system isn't as good at whisking away the fumes and smoke as a spring breeze can be when you're welding outside.

As you can imagine, the biggest challenge presented by outdoor welding is the elements. The weather can wreak havoc on the quality of your work and the integrity of your equipment. Precipitation can make welding unnecessarily dangerous, and forget welding successfully when the temperature is below 40 degrees Fahrenheit if you don't preheat the materials. And even sunshine, which is usually a pleasant part of welding outside, can pose a problem: Sometimes direct sunlight makes certain types of welding helmets function poorly. (Read more about welding helmets in Chapter 3.) Finally, if you're welding outside where other people will be present, you really need to set up a safety screen so your audience's eyes aren't harmed by the welding process if they happen to look at it.

Equipping Your Welding Shop

You'd be hard pressed to find a skilled trade or craft that doesn't require some specialized equipment, and welding is definitely no exception. But hey, you're talking about joining metals, not making ham sandwiches, so I'm sure it comes as no surprise that you need some specific gear in order to get the job done right. You can get a look at some of the necessary equipment in Figure 4-1.

In this section, I let you know what kind of general equipment you need for a welding project. Keep in mind that plenty of other types of gear are necessary for certain kinds of welding, and I tell you about those tools and materials in the parts of this book that dig into the details of specific welding types. (For example, you need certain items for flux core arc welding that you don't need for the other kinds of welding, and I cover those specific items in Chapter 9.)

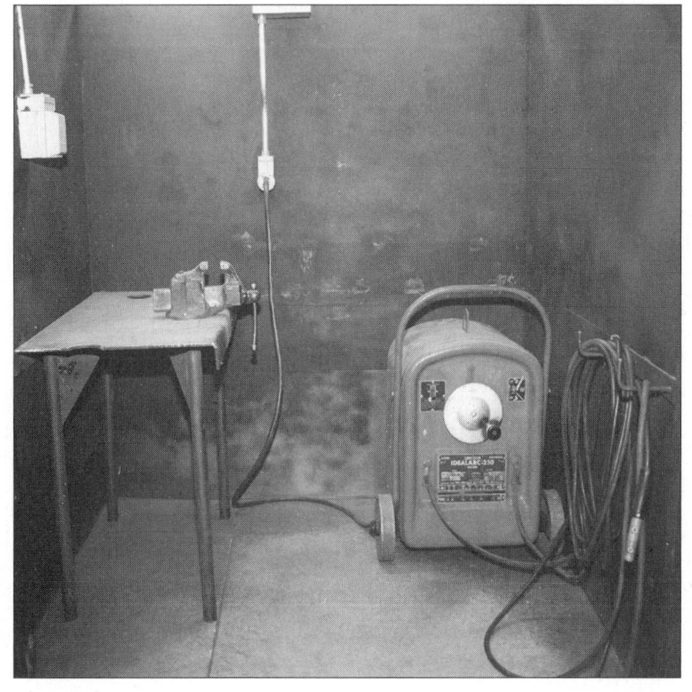

Figure 4-1: A well-equipped welding area setup, this one for stick welding.

The most important welding equipment you'll ever own is your safety equipment. I don't cover that information in this section, so be sure to flip to Chapter 3 so that you understand everything you need to stay safe and healthy as you weld.

You have plenty of options when you're looking to obtain your welding equipment. A lot of different manufacturers are producing tools and materials now; if you do a quick Internet search for a chipping hammer — one of the most basic welding tools — you find dozens of options from many different suppliers in a range of prices.

So which equipment is right for you? In many cases, you can let your budget be your guide; look around (at hardware or home improvement stores, on the Internet, in welding catalogs, and so on) at the various options and find some that fit both your needs and your price range. Then, to narrow your choices, read some online reviews of the various products and talk to knowledgeable salespeople at the retailers you're considering. If you have friends, family members, or co-workers who weld, pick their brains about the tools and materials they use.

If you're looking to buy some welding equipment and you're worried that your budget isn't going to cover all the items you need, don't be afraid to investigate used options. You can find good quality used equipment from online retailers and also at auctions.

Making sure you have the basic hand tools

Some of the tools you need for your welding projects are basic hand tools. Several of these items are pretty common, so you may very well already have one or a few of them in your toolbox now. If you don't, though, you're going to have to get your hands on them, because you can't do without them.

With hand tools used in welding, try to stick to metal (not wood or plastic) construction because of the heat and sparks involved. Metal tools hold up and perform better over the long run.

Following are the basic tools you need, some of which are shown in Figure 4-2:

- **Ball-peen hammer:** *Ball-peen hammers* are used for striking chisels and punches. They're also handy for straightening metal. The rounded head is used for *peening*, which relieves the stress in a joint that you've just welded. You can pick up a good ball-peen hammer at a hardware store for around $15.
- **Chipping hammer:** A *chipping hammer* is a specialty hammer you use to remove excess material (called *slag*) from your welds. Chipping hammers come in a range of sizes, but they're typically a foot long and weigh about a pound. If you're buying a new chipping hammer, you can expect to spend between $10 and $30.
- **Crescent wrench (12 inches):** Keeping a 12-inch crescent wrench on hand when you're welding is always a good idea. It fits the majority of the nut and bolt sizes you encounter in welding.
- **Level:** A level is necessary to make sure the pieces you're welding are at the correct horizontal or vertical position and also to test the accuracy of completed projects. Make sure you have an extremely durable level because they take a lot of abuse.

- **Pliers:** You definitely need a good pair of pliers for welding, and the run-of-the-mill pliers most people have in their toolboxes don't really get the job done. You need *slip joint pliers* (pliers that let you increase the range of their jaws) with compound lever action to handle hot metal. You can also use tongs. Depending on the size of the pliers and the number of bells and whistles involved, you can expect to spend anywhere from $7 to more than $20.

Pliers or tongs are great for picking up hot metal. Grabbing pieces of hot metal while wearing your leather welding gloves can be dangerous, but it also hardens the leather, making the gloves less effective (and less comfortable) over time.

- **Soapstone or silver pencils:** You can't use regular pencils or pens to write on metal, so you need to use soapstone or silver pencils. You can find both at your local hardware store for only a few dollars.
- **Steel square:** The *steel square* is a durable tool that you can use to keep your project in square during assembly. It's essentially a right angle made out of steel, with one side 16 inches long and the other side 24 inches long. The face side of a steel square has a ruler that marks whole, half, quarter, and eighth inches. The back side is divided into whole inches, halves, and quarters.
- **Straightedge:** A straightedge is a must for drawing straight lines, transferring measurements, and double-checking for straightness. After all, nobody wants a crooked weld.
- **Tape measure:** You have to be able to measure and mark your materials to cut at the right dimensions, so a tape measure is a must. Luckily, tape measures are one of the most common tools out there. You can handle most jobs with a 25-foot-long, 1-inch-wide tape measure.
- **Vise grips:** These wrenches are combination tools that function as pliers, wrenches, portable vises, and clamps. They aren't intended to replace box-end wrenches because of the damage they do to the fittings you're working on. Most people prefer to also have a pair of pliers (as mentioned earlier in this list) in addition to vise grips.

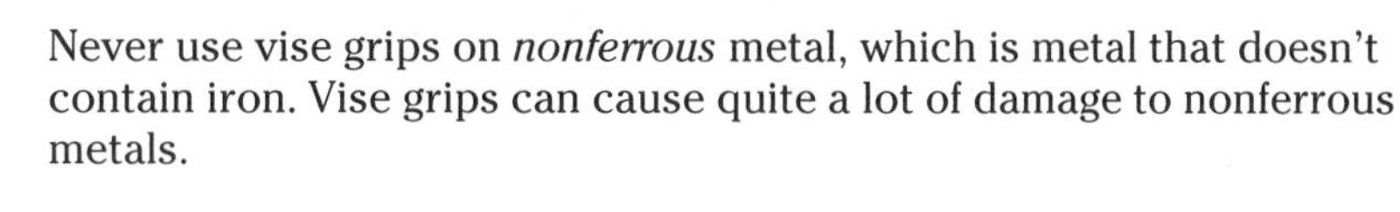
Never use vise grips on *nonferrous* metal, which is metal that doesn't contain iron. Vise grips can cause quite a lot of damage to nonferrous metals.

- **Wire brush:** You use wire brushes to clean up welds after you've chipped away at them. I recommend getting a brush with a nice long handle. If you want to kill two birds with one stone, some manufacturers sell a combination chipping hammer/wire brush. You can get a good wire brush for $10 to $20.

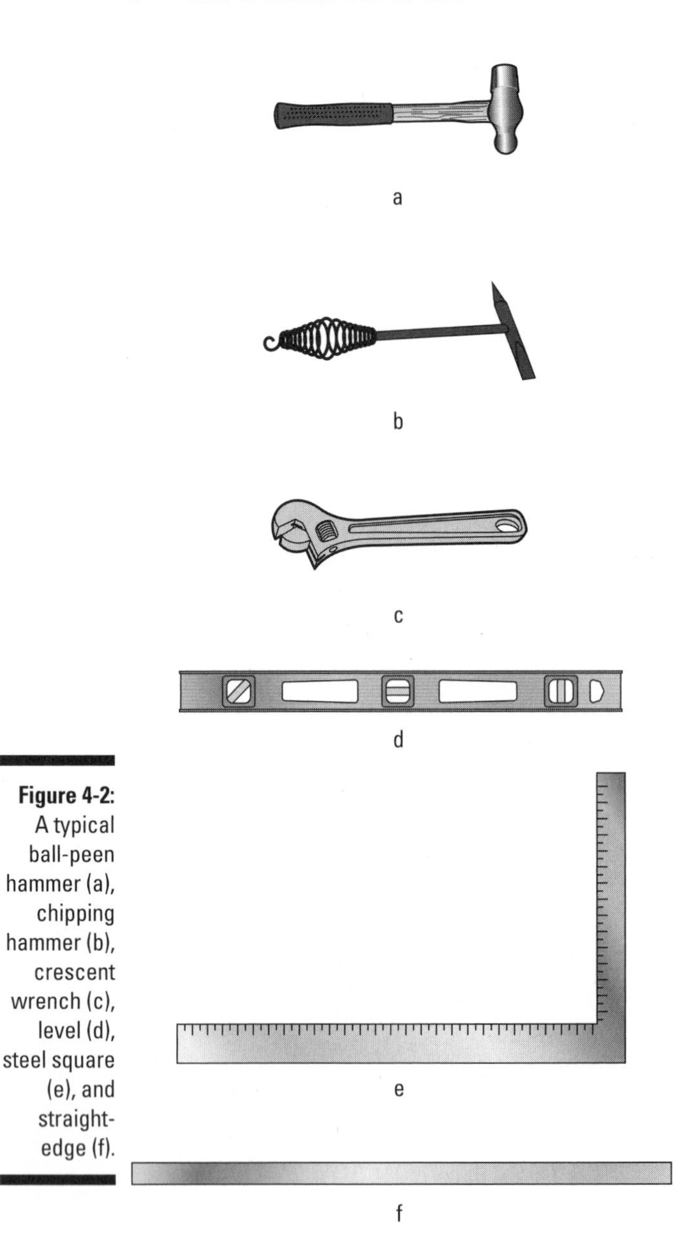

Figure 4-2: A typical ball-peen hammer (a), chipping hammer (b), crescent wrench (c), level (d), steel square (e), and straight-edge (f).

Choosing a welding table

You need a very sturdy, reliable surface to use as a foundation for welding practice as well as many of your welding projects, especially the smaller jobs. The best option is a welding table. Don't try to use a normal table or work bench, because the wood or metal construction will soon be spattered with (and probably burned by) welding materials.

A good welding table includes a steel top and angle iron or sturdy pipe for legs. (Check out the welding table in Figure 4-3 to get a feel for what one looks like.) Your local welding shop can build you a welding table, or you can buy one at a hardware or home improvement store. The table size you need is determined by the size of the projects you plan to work on. I recommend getting a table that is ⅜ inch thick, with a top that is 36 inches by 48 inches. Expect to spend at least $50, and remember that after you get the hang of welding, you can make your second table instead of buying it.

Figure 4-3: A typical welding table.

Be sure that you can adjust the height of your welding table, or be certain before you buy it that you can be comfortable working at the table. You don't want to have to bend down or stretch to use the table.

Selecting your welding machine

One of the most important pieces of welding equipment you buy is your *welding machine,* the apparatus that provides the energy you use to make your welds and complete your welding projects. The machines come in many types, sizes, and prices to fit the needs of any welder. You can find inexpensive machines for small shops and projects, and expensive, more powerful welding machines for big shops and huge projects.

One of the most important considerations you need to keep in mind when you're deciding which welding machine to acquire is the type of welding you plan on doing. Welding machines are available for all the various types of welding: stick welding, mig, tig, and more. (You can read more about the various types of welding in Chapter 1, and in the chapters that make up Parts II, III, and IV, where I dive into the details of each welding type.)

Welding machine basics

Following are a few welding machine considerations that you can keep in mind when you start thinking about the type of machine you want to buy for your shop.

- **Power source:** Be sure to purchase a welding machine that fits the electricity power source that you have available. Most household or small shops have only single phase electricity available.
- **Current:** Welding machines are also sometimes classified by the type of current they require. Some machines use *alternating current* (AC), which is an electrical current that flows back and forth at regular intervals in a circuit. Other welding machines use *direct current* (DC), which flows in one direction.
- **Duty cycle:** Welding machines are rated on *duty cycles*, which indicate how many minutes out of ten the machine can run continuously at its maximum amperage output. (Most welding machines have an output of at least 200 amps.) A welding machine's duty cycle is based on the manufacturer's recommendation. For example, if a 200-amp machine is rated with a 60-percent duty cycle, you can use it continuously (and safely) for six out of ten minutes when you have it set to produce an output of 200 amps. You can get a *constant potential* (CP) welding machine that maintains a constant flow of electricity, which results in a 100-percent duty cycle.

- **Material type and size:** Some welding machines can weld just a couple of different types of metal, and others can weld a huge range. You should also keep the size (thickness) of the metals you're welding in mind. If you're welding materials that are less than 3⁄18 inch thick, most welding machines are powerful enough for you. If you're welding thicker metals, you may want to get a machine that can produce more amperage.

New or used: Which welding machine is right for you?

After you understand the different types and classifications of welding machines (see the preceding section), you need to decide whether to buy a new machine or one that has been used.

The advantages and disadvantages afforded by new and used welding machines are similar to what you'd expect for most other types of machinery or equipment. New welding machines have no wear and tear, and you can often buy one with a warranty that helps protect you if the machine (or part of it) happens to break down soon after you buy it. Plus, new machines come with instructions that can be extremely helpful for a beginning welder, whereas used machines often no longer have the directions. On the downside, new machines are almost always more expensive than comparable used models. New stick welding machines typically cost at least $400, and mig welding machines are often twice that amount. A brand new tig welding machine can run you $1,200 or even more, depending on the size and features of the machine.

If you want to buy a new welding machine, you can find them at larger hardware and home improvement stores, welding supply stores, and through online retailers.

Used welding machines can be a good route for people who are just starting out with welding and don't want to sink a lot of money in a new machine. You can find quality used welding machines for half the cost of a new machine. The good news is that most welding machines don't wear out or become ineffective unless they're mistreated or maintained poorly.

If you want to buy a used welding machine, you likely have three options:

- **Auctions:** Equipment or general auctions in your area can be a great source for used welding machines. Call ahead or look through auction listings to see whether any welding machines are up for bidding.
- **Welding supply stores:** Many welding supply stores offer incentives for welders to trade in their used welding machines when they're looking to buy a new machine. The stores then sell the used machines, sometimes after carrying out some basic maintenance or making any necessary repairs to the machine.

- **Individual sellers:** If you have friends or family members who weld, you may ask them if they have a used welding machine that they're interested in selling in the near future. That's a good situation because it means you know the seller and are probably familiar with how he treats his equipment. (Although if that relative is your reckless Uncle Bob, you may want to think twice.) You can also find individual sellers looking to sell used machines in classified ads or through online listings.

If you're interested in buying a used welding machine, be sure to require the seller to turn on the machine and produce a *weld bead* (deposit of filler metal from a single welding pass) for you. That basic operation check can help you avoid buying a worn-out or broken machine.

I know it seems obvious, but I'm always surprised at the number of people who buy a welding machine only to get it back to their shop and realize that the cables are too short for the projects they plan to work on. Be sure to check a welding machine's cable length before you make a purchase!

Stick welding machines

Stick welding is one of the most widely practiced welding techniques. (You can read all about stick welding in Chapters 5 and 6.) One of the biggest reasons for stick welding's popularity is the fact that it's inexpensive — in fact, the equipment used in stick welding is the least expensive of all the electric welding processes. Stick welding machines include the machine itself, as well as the welding leads, ground clamp, and electrode holder (see Figure 4-4). That's all! If you're going to be doing any stick welding and you want to get your hands on a stick welding machine, the big choice you have to make is whether you want an AC or DC machine.

- **DC:** DC stick welding machines give you a continuous flow of electricity, and they make it easier to produce good welds because they give you a higher level of control.
- **AC:** AC stick welding machines have one very important advantage: price. AC stick welding machines are the cheapest welding machines you can buy, and they require little or no maintenance. However, they can be a little harder to work with than DC machines, especially for new welders.

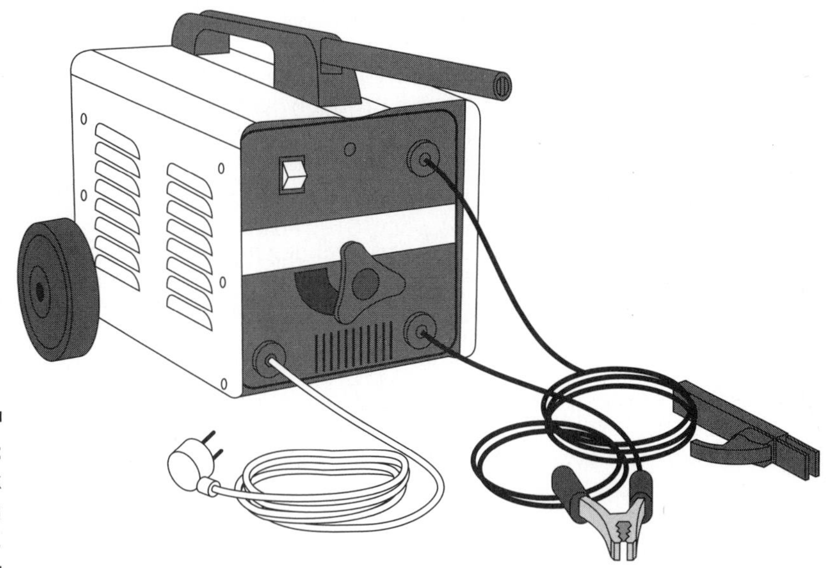

Figure 4-4: Typical stick welding machine.

Mig welding machines

Mig welding machines are more complex than stick welding machines and include a welding gun, a spool of electrode, an electrode feeding unit, a power supply, cables, hoses, and a cylinder of shielding gas. (Flip over to Chapters 9 and 10 for more on those terms and other mig welding details.) You can check out mig welders at your local welding supply store, and you can also get parts (like refills for your shielding gas cylinders) there as well.

Mig welding machines (see Figure 4-5) are designed to weld a range of materials, including steel, stainless steel, and aluminum, to name just a few. However, if you want your mig welder to work on a variety of metals, you have to buy a few attachments. A mig welding machine without any additional attachments can weld steel, but if you want to weld aluminum, for example, you need a spool gun. Want to weld stainless steel? You need to change your shielding gas cylinder. Your local welding supply store can tell you all about what you need to change on your mig welder if you want to take advantage of the machine's versatility and work on many different metals.

Like the other types of welding machines, some mig welders are more powerful than others. The power you need is determined by the thickness of the metals you plan to weld. If you're going to be welding materials that are more than ⅜ inch thick, consider buying a more powerful mig welding machine.

Some manufacturers are now producing mig welding machines that are self-setting. You just have to dial in the thickness of the material you're welding and the machine sets the amperage level on its own. Pretty slick!

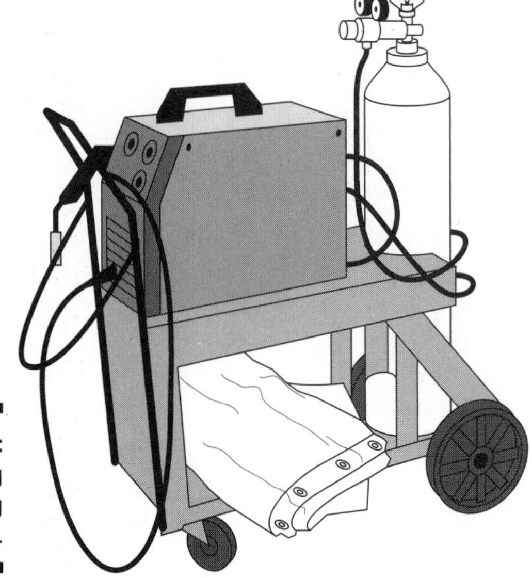

Figure 4-5: A mig welding machine.

Tig welding machines

Tig welding machines are the most versatile of all the welding machines. With a tig machine (see Figure 4-6), you can weld all ferrous and nonferrous metals. Most tig machines allow you to change *polarity* (the direction in which the electricity flows), and that allows you to weld a wider range of materials.

Tig welding machines and stick welding machines are somewhat interchangeable with a little modification. Stick welding machines can be used for tig welding if you buy and add on a number of attachments, and vice versa. If you're planning on working on both tig welding and stick welding projects, I suggest buying a tig welding machine and also purchasing the attachments — a stick welding electrode holder with a lead and plug on the end that matches the machine — that allow you to use it for stick welding. It's easier to make a stick welding machine from a tig machine than the other way around.

If you want to read more about the ins and outs of tig welding machines, check out Chapters 7 and 8.

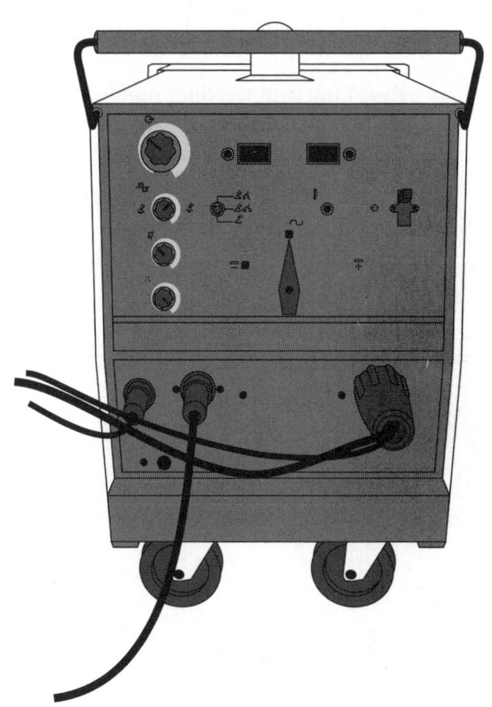

Figure 4-6: A tig welding machine.

Considering a few accessories for your welding shop

As with most types of equipment, you can add a few bells and whistles to your machine in the form of accessories. These items can make your machine easier to use, so you may consider whether one or more of the following are right for your setup:

- **Cable holder:** A hook mounted to the side of the machine that you can use to hold your lead and ground cables, cable holders help keep your work area more organized (and safer).
- **Cup holders:** I know it's hard to believe, but you can now get cup holders that mount to your welding machine. Just make sure your drink has a lid.
- **Storage:** You can get small drawers for some welding machines now that can helpfully store the various small parts that you need when welding.

- **Welding machine carriage:** These carts have wheels that make moving your welding machine easier. They're almost always made of steel, and you can buy one for about $200. That seems like a lot, but some welding machines weigh more than 100 pounds, and spending $200 for a cart is much cheaper than spending twice or three times that amount at the chiropractor's office!

These accessories aren't really necessary for you to own, but they are handy things to have. For information on even more tools and accessories that you may want for your welding shop, see Chapter 19.

In addition to the basic tools, welding table, welding machine, and welding machine accessories I discuss in the preceding sections, this section lists several other tools and pieces of gear you may want to buy and have on hand when you're getting ready for a welding project or setting up your shop. You can find all these items at welding supply stores or hardware/home improvement stores.

- **Air compressor:** If you're going to be using any pneumatic tools, you need a good air compressor. Cost: $160.
- **Anvil:** An anvil is a good tool to have around when you need to hammer or straighten out a large piece of metal. (It's also useful, of course, if you're a coyote and you're trying to kill a roadrunner.) Keep in mind that you can also use a large, solid, secured piece of metal (a railroad rail, for example) in your shop in place of an anvil. Just don't go pulling up train tracks to save money. Cost: $150.
- **Creeper:** A *creeper* (shown in Figure 4-7) is a small platform with wheels that you can lie down on and use to roll around. Creepers are helpful when you need to get up under something. Cost: $30.

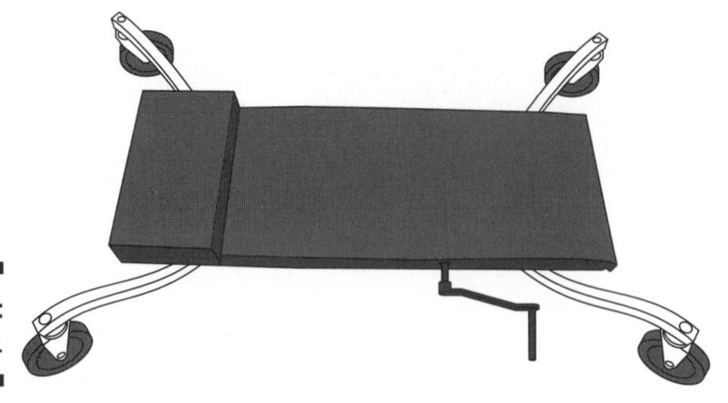

Figure 4-7: A creeper.

- **Drill press:** I think having a good, small drill press on hand in the welding shop is extremely handy. Be sure you have plenty of good bits that fit the press. Cost: $100 to $1,000.

  Don't buy a drill press that's too large or expensive for your needs. Bigger isn't always better.

- **Grinder:** *Grinders* are useful for grinding down rough surfaces. If you get a grinder, make sure to also pick up some ferrous and nonferrous grinding discs that fit the machine. Cost: $85 to $125.
- **Hacksaw:** Hacksaws are extremely useful for cutting metal. Cost: $8 (includes a package of blades).
- **Portable welding curtain:** These items (see Figure 4-8) are usually made of vinyl, and they can help to shield bystanders from the dangerous rays created while welding. Cost: $10.
- **Vise:** Consider mounting a small vise to your welding table. It can be very helpful when you're trying to work on small projects that are tough to manipulate while welding. Cost: $25.
- **Wall clock:** It may sound simple, but having a clock mounted on the wall is a much easier way to keep track of the time while welding than having to take off your gloves and push up your sleeves to check your watch. Welding is fun, and time flies when you're having fun, so why not have an easy way to keep track of it? Cost: $3.

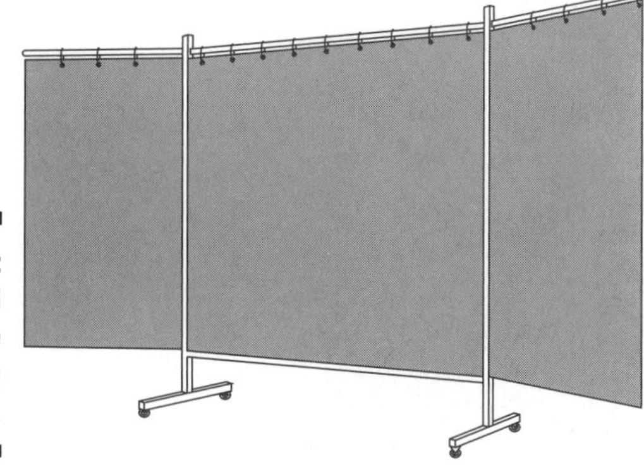

Figure 4-8: A standard portable welding curtain.

Part II

Welding on a Budget: Stick and Tig Welding

"Sure, you can borrow some of Grandma's things for your school project. There's an empty pickle jar, some baking soda, and let's see, where's that old TIG welder of mine? Here it is, in the pantry."

In this part . . .

If you stop someone on the street and ask them to name a type of welding, they may sock you in the nose. They may also say "stick welding," because that's the most common type of welding done in the world today. It's an arc welding process that's easy to get into, and one that can really serve you well if you take the time to understand and practice it.

Tig welding, on the other hand, isn't quite as well known as stick welding. But it's still very widely practiced, and it's definitely something you want to consider picking up and adding to your welding repertoire.

I cover both stick and tig welding in this part, so if one of those two processes is high on your list of welding techniques to explore, this part is for you.

Chapter 5

Getting on the Stick: Understanding Stick Welding

In This Chapter

- Considering stick welding's advantages and disadvantages
- Figuring out when you should use stick welding rather than other processes
- Knowing important stick welding factors
- Perusing the equipment you need for successful stick welding

No matter what level of welding you consider — from industrial and commercial applications to tinkering around with a small project or two on the weekends — stick welding is always one of the most prominent and dynamic welding techniques, partially because it's one of the more inexpensive welding types.

Regardless of what you want to do with your welding skills, you need to understand the basics of stick welding so that you can make it a key aspect of your welding repertoire. In this chapter, I offer an overview of many critical stick welding basics.

Looking at the Pros and Cons of Stick Welding

As with any welding technique, stick welding offers distinct advantages and disadvantages. The beauty of stick welding (also known as *shielded metal arc welding* or SMAW) is its versatility and portability. Using stick welding you can weld metals in a wide variety of thicknesses and from a range of positions (In Chapter 6, I go into detail about how you can stick weld in multiple positions.)

You can also stick weld just about any kind of metal, including steel, stainless steel, aluminum, and just about anything in between. If you're just getting started with welding as a hobby and you know that you're going to want to weld a lot of different materials, stick welding is a solid choice for you.

Following are a few more characteristics that have helped make stick welding one of the most commonly used welding processes.

- Stick welding equipment is less expensive than the equipment you need for any other welding process, in some cases costing half or a third as much.
- Stick welding equipment is lightweight, and it isn't as complex as the gear you need for other welding processes.
- A range of stick welding electrodes allow you to tackle many different welding applications.
- Stick welding allows you to weld in confined spaces.
- Stick welding operations are very portable. If you know that you're going to have to weld outside of the friendly confines of your welding shop, stick welding is probably a good choice.
- You don't have to prepare the surfaces of the metal you plan to weld as much as what you'd have to do to get the metal ready for other welding processes.

Stick welding sounds great, doesn't it? It really is a great technique, but it's not without flaws or drawbacks. Here are a few of the cons associated with stick welding.

- Stick welding can be very labor intensive, which can result in high labor costs.
- You always waste part of your stick-welding electrodes when you're working on a stick-welding project because you have to stop using an electrode when two inches of it remain. (More about that in Chapter 6.)
- Stick welding requires that you do a lot of cleanup as you work your way through a weld. You must clean the weld after every pass and after every pause or stop.

- Stick welding produces more heat sparks and spatter than any other welding process. Be sure you have the necessary personal protective equipment and you're familiar with welding safety precautions before you get started with stick welding. (You can check out Chapter 3 for tons of great safety information.)
- As compared to the other welding processes, stick welding is the least efficient in terms of converting electricity to useful welding heat.
- The welds that result from stick welding often have to be chipped and ground down, which can create harmful dust.

- The chipping and cleaning necessary with stick welding can be extremely noisy, and that poses a risk for hearing damage.
- Although stick welding equipment is pretty simple, the technique you need to master to stick weld successfully can be difficult to pick up.
- Stick welding electrode stubs are short and round, and they can cause a slipping hazard if you accidentally step on one.
- Stick welding can be slow because you have to stop welding when you work your way through an electrode. You also have to stop often to chip away at accumulated slag on your welds.

Understanding the Factors That Influence Stick Welding

When you're stick welding, you can control several key factors that determine how your weld comes out. These welding variables allow you to change the finished product of the weld, including the height and width of your welding bead, the amount of spatter, and the penetration. (*Penetration* is how deep the weld goes below the surface of the metal you're welding.) I find that new welders really benefit by thinking about these factors before they dig in to the details of stick welding. Here are some pointers that may help improve your welds.

- You can increase or decrease the amperage of your weld by using the controls on your stick welding machine.
- You can increase or decrease the *speed of travel* (the speed at which you move the electrode along the weld joint) of your weld.
- You can increase or decrease the *arc length* (the distance from the end of the electrode to the metal you're welding) of your weld. Your arc length should be the same distance as the diameter of the electrode core wire you're using.
- You can change the angle at which you hold the electrode to affect penetration. If you hold the electrode so it's basically perpendicular (at a 90-degree angle) to the parent metal, you get maximum penetration. If you hold it at a tighter angle — 45 degrees, for example — you get less penetration.
- You can use a side-to-side motion with the electrode to control the width of your weld.
- You can use a variety of electrode types and sizes. Check out Chapter 6 for details.

- You can use different kinds of *polarity* (the direction electricity flows in direct current or DC). Check out the "Polarity" section later in the chapter for more on this topic.

If you control these factors correctly, you end up with a good, solid stick weld. If you can't, you end up with a poor, weak weld. You can see examples of both in Chapter 6.

Getting Familiar with Stick Welding Equipment

As I mention earlier in this chapter, stick welding equipment is much less expensive than the equipment you need to perform any other welding process. That said, you still need to get and understand several pieces of equipment and a number of tools before you can get started with stick welding. That's what this section is all about.

Discovering the differences among stick welding machines

The foundation of your stick welding operation is your stick welding machine. The machine provides the electricity that creates the welding arc. Although stick welding machines are simpler than the machines you use for the other welding processes, as a category they still have plenty of variety and a number of differences from one machine to another. The following sections fill you in on some of the variables that you need to keep in mind as you decide which kind of stick welding machine to buy or use for your projects, and you can see what a typical stick welding machine looks like in Chapter 4.

Polarity

You use either alternating current (AC) or direct current (DC) electricity to stick weld. Some machines provide one of the two types of current, and other machines provide both.

DC machines typically have a control that allows you to choose either *direct current electrode positive* (DC+) or *direct current electrode negative* (DC–). When you have the machine set on DC+, electricity flows from the metal piece you're welding to the electrode you're using. With DC–, the electricity flows from the electrode to the metal. I recommend using DC+ whenever possible.

AC machines provide alternating current only, and they're the least expensive of the stick welding machines. Stick welding machines that allow you to toggle between AC, DC–, and DC+ are by far the most versatile, but they're also more expensive.

Output rating and duty cycle

The *output rating* of a stick welding machine is the amount of amperage it can put out. The higher the output rating, the more powerful the welding machine — and you need a powerful machine if you're interested in stick welding thick pieces of metal.

The duty cycle of a stick welding machine is related to its output rating. *Duty cycle* is expressed as a percentage that tells you how many minutes out of ten you can use the welding machine at its maximum amperage output. So if you have a stick welding machine that can put out 200 amps and its duty cycle rating is 60 percent, that means you can run it at 200 amps for six minutes out of ten. As you can imagine, stick welding machines with higher duty cycle ratings are more expensive.

Power source

Stick welding machines can have one of three different types of power source: generators, transformers, or inverters.

- **Generators (DC generators)** were the first type of power source for stick welding. You don't plug a generator stick welding machine (shown in Figure 5-1) into the wall; it provides its own power. The generator is driven by gasoline, diesel, or natural gas, and it's still a popular power source, especially for stick welding in the field, when you need to weld but have to provide your own source of electricity. In fact, some generator stick welding machines are mounted on trailers so that they're more portable.
- **Transformer** machines don't create their own source of electricity (like generator machines); instead, you have to plug them in. You can adjust the amperage of these machines in a number of ways, and the machines are very efficient; however, transformer machines provide AC current only, which means you're somewhat limited in the kinds of electrodes you can use. Many transformer stick welding machines are being replaced with DC-compatible welding power sources (generators or inverters). Figure 5-2 shows a typical transformer *buzz box*, a slang term for an AC transformer-type welder. As you can guess, the machine makes a buzzing sound when you turn it on.

Figure 5-1: A generator stick welding machine.

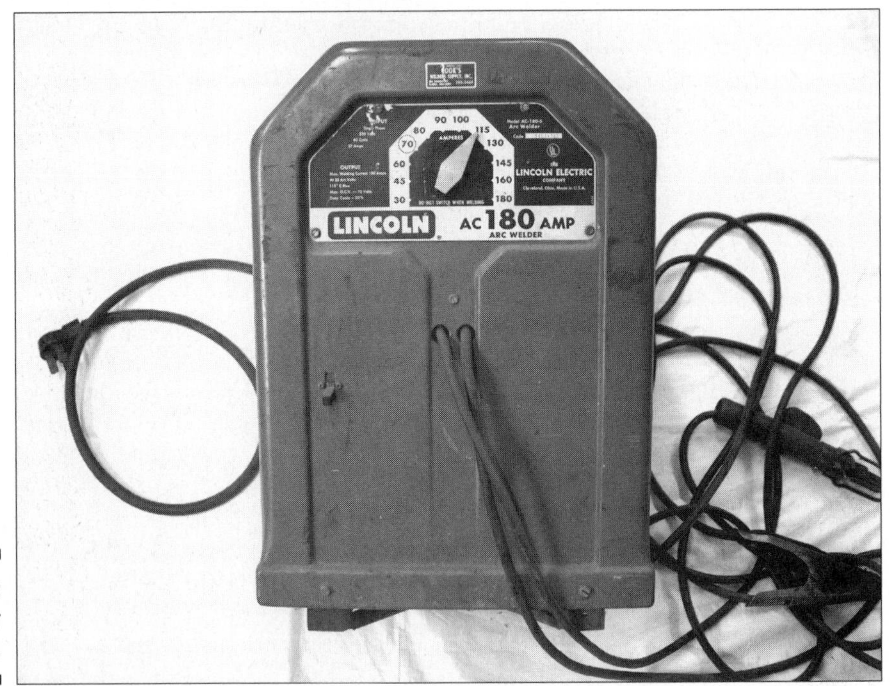

Figure 5-2: Transformer buzz box.

- **Inverter** machines are relatively new on the scene: They were invented in the latter half of the 20th century. They're more complex and advanced than generator or transformer machines, and as a result they're becoming more and more popular. I won't go into the gory electrical engineering details of how these machines work, but just know that the technology allows for very powerful machines that don't weigh nearly as much as generator or transformer machines that produce similar amounts of amperage. Inverter machines also offer both AC and DC, reduced fumes, improved penetration, and reduced spatter.

You can see an inverter stick welding machine in Figure 5-3.

Figure 5-3: An inverter stick welding machine.

Setting up your stick welding machine

After you have a stick welding machine, you have to get it set up in your shop. Put it close enough to where you plan to weld that the cables can reach but far enough that the machine won't be constantly hit with sparks. Stick welding machines are air cooled, and you don't want sparks and debris to be floating around in the air that gets sucked in to cool the machine's parts. Try to avoid running the cables across areas of the floor where you (or others) will be walking, and never tie your cables to ladders or any other equipment in the welding area.

Don't put the machine near any containers that hold flammable liquids or gases, and make sure there aren't any water leaks near the machine. If your stick welding machine plugs into a wall outlet, be sure you know which breaker or switch on your electrical panel matches up with that wall outlet, so you can cut off power to the stick welding machine from the panel in the event of an emergency.

Nailing down the basics of stick welding's electrodes

The electrodes used for stick welding have a pretty basic construction. They have a metal core, which can be made of lots of different types of metal, and an outer covering of *flux*. The flux coating shields the arc from the air and from potential contaminants, thereby maintaining the strength and integrity of the finished weld. Flux helps to eliminate metal *porosity* (tiny holes in the weld), cracking, undercutting, and spatter. It also allows you to store stick welding electrodes for long periods of time.

Chapter 6 gives you the details on how to choose stick welding electrodes to suit the type of project you want to work on, but here I give you some general rules to bear in mind when it comes to using and maintaining your electrodes.

- **Keep your electrodes dry.** Allowing them to be exposed to water or moisture can ruin them. Wet electrodes can cause porous, weak finished welds.

  You can tell whether the electrode you're using has been exposed to water or moisture because wet electrodes give off a crackling sound when you use them. Dry electrodes give off a frying or sizzling sound.

  You can heat and dry out some electrodes if they become wet, and several manufacturers produce specialty ovens to do just that. (Don't put electrodes in your kitchen oven, or your next meatloaf will come out tasting kind of funny.) To be honest, though, an electrode oven is probably a little more equipment than you need if you're just welding as a hobby; you're better off just taking special care to keep your electrodes dry.

- **Handle your electrodes with care.** Rough handling can cause the coating to crack loose from the core wire, rendering the electrode unusable.

  Don't use an electrode if the core wire is exposed because the missing flux will cause welding defects.

- **Avoid brands that burn off too fast on one side.** That problem is more common when you're using DC, and it's usually caused when the flux coating isn't completely centered on the wire. (It's a manufacturing error.) If it happens to you while you're welding, rotate the electrode 180 degrees in the electrode holder and continue to weld. (You can grab it with a gloved hand with the machine on while the electrode is still hot.) If the problem persists, don't buy that brand of electrode again.
- **Watch out for arc blow.** *Arc blow* occurs when your welding arc blows out in a direction that isn't consistent with the direction you're trying to weld. It's most common when you're trying to weld a corner, or at the end of a weld. The easiest way to correct the problem is to weld toward your ground and keep your arc length nice and short. I dive into the details of striking and maintaining an arc for stick welding in Chapter 6.

Choosing tools and supplies every stick welder needs

You need plenty of supplies and tools for every welding process; read all about those in Chapter 4. You can also get familiar with the safety gear necessary for welding in Chapter 3. But you really should have some other items if you plan to stick weld, and that's what I cover in this section.

- **Coveralls:** Coveralls for stick welding should be heavy enough to prevent the ultraviolet arc rays from penetrating the fabric and burning your skin. Don't wear anything with cuffs, and make sure any pockets are covered with a flap that buttons or snaps so that flying sparks don't get trapped in them.
- **An auto-darkening or appropriately shaded welding helmet:** You can brush up on welding helmet basics in Chapter 4; review Table 5-1 to make sure your helmet lens is dark enough for the amperage you're using.

  You always want to use the darkest shade and yet still be able to see the work to lessen the chances of eye damage.

Table 5-1 Welding Helmet Shade Guidelines for Stick Welding

Amperage Range	*Suggested Shade*
60 to 80	8
70 to 90	9
80 to 120	10
100 to 140	11
120 to 160	12

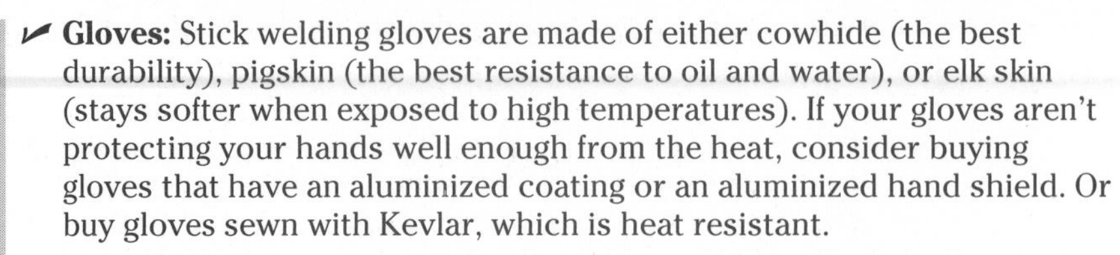

- **Gloves:** Stick welding gloves are made of either cowhide (the best durability), pigskin (the best resistance to oil and water), or elk skin (stays softer when exposed to high temperatures). If your gloves aren't protecting your hands well enough from the heat, consider buying gloves that have an aluminized coating or an aluminized hand shield. Or buy gloves sewn with Kevlar, which is heat resistant.

  If you're stick welding and a molten piece of metal gets in the seam of your glove, leave it alone! If you try to pull it out, you'll probably remove the threads that hold your glove together.

- **A respirator:** If your welding area isn't extremely well ventilated, you need a respirator to protect your lungs. I discuss respirators in Chapter 4.
- **An appropriate electrode holder:** You want an electrode holder that's fully insulated and has spring or twist-type jaws. The electrode holder in Figure 5-4 is a good example. If your electrode holder isn't fully insulated, you'll probably get shocked; if it doesn't have the right kind of jaws, they won't grip the electrode properly.

  Make sure the electrode holder you use is rated for the amperage you plan to weld with.

- **Ground clamps:** *Ground clamps* (see Figure 5-4) are designed to ground your welding table so that the piece you're working on conducts the electricity well, which ensures a good finished weld.

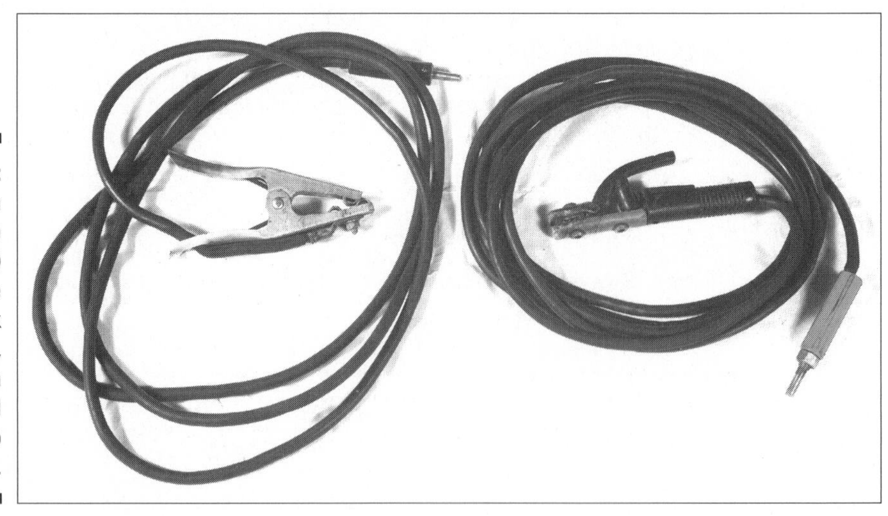

Figure 5-4: An electrode holder (left) appropriate for stick welding, along with ground clamp (right).

- **Properly sized and insulated welding cables:** In addition to cable length (which I discuss in "Setting up your stick welding machine" earlier in the chapter), you want to ensure your welding cables are large enough for the job. If your cables aren't large enough to handle the stick welding amperage you want to use, they'll overheat and fail. Most cables are made out of stranded copper wire, and you should aim to buy cables that are very well insulated to help avoid overheating.
- **A combination square:** A *combination square* is different from a steel square I cover in Chapter 4 because you can make the head slide along the blade and clamp it at any desired measurement, which is a useful feature when you're stick welding. You can also pull out the blade and use it as a ruler.
- **Soapstones:** Soapstones are made primarily of talc, and they're useful for drawing lines, shapes, and marks on a huge range of metals. Stick welding allows you to weld just about any kind of metal, so a soapstone is very useful. You can sharpen it with a file or grinder. You can also buy a soapstone holder to put your soapstone in so that you can use it like you would a pencil.

Figure 5-5 illustrates the combination square and soapstone. If you're all geared up with the necessary stick welding supplies, you're ready to get started.

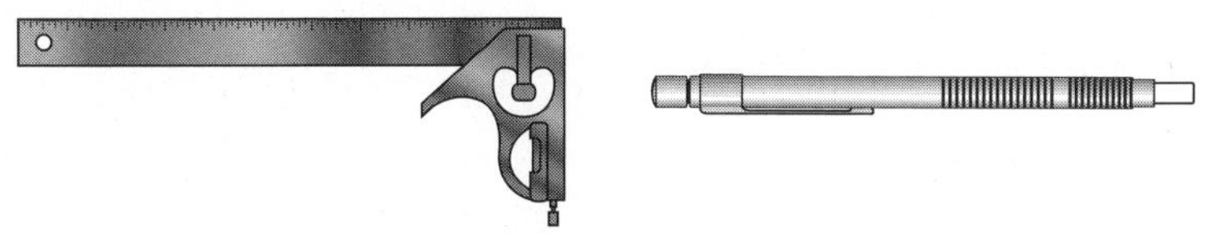

Figure 5-5: A combination square and soapstone.

Chapter 6

Getting to Work with Stick Welding

In This Chapter

- Getting ready to stick weld
- Making sure you have your stick welding machine set properly
- Figuring out how to strike and sustain an arc
- Understanding how to stick weld in all positions

Stick welding (also known as *shielded metal arc welding* or SMAW) has become one of the most widely used welding processes for repairing metal. When you stick weld, the intense heat of the arc melts the end of the electrode and the base metal you're trying to weld. The melting electrode goes across the arc and deposits itself in the molten pool of metal to form the weld bead. The electrode also has a coating, which melts to form a gas that shields the molten pool of metal from the impurities in the air. It also forms *slag* to protect the cooling weld.

As you can read in Chapter 5, stick welding has become increasingly popular because it's inexpensive and relatively versatile. If you want to get started in welding, you really do need to know how to stick weld, which I show you in this chapter. I start with some basic information beginning a stick welding project and then take you through the process of electrode selection — a very important step when stick welding. I also provide instructions on how you can stick weld in a range of different positions.

Preparing to Stick Weld

Before you dive into a stick welding project, make sure stick welding is the most appropriate welding technique for the job at hand. If you're not sure, flip to Chapter 5 to read through the different scenarios that call for stick welding. You don't want to get all set up for stick welding (and maybe even get started with the project) only to realize that mig would've been the better choice all along.

If you're completely confident that stick is the best route for a given job, you then need to get your work space prepped, gather your tools, and make some smart decisions about the kind of electrodes that you need for the task ahead of you. I cover all those considerations (and more!) in the following sections.

As with any type of welding preparation, getting ready to stick weld requires that you take a number of safety precautions. If you haven't already, be sure to check out Chapter 3 to read up on how to keep your welding project as safe as possible.

Setting up your work area

Stick welding requires that you keep an extremely tidy work space. Out-of-place tools and other items only make a stick welding project harder and more likely to take longer than it needs to. Obstructions and clutter also contribute to generally unsafe working conditions, and when you're using large amounts of electricity to melt metal, why put yourself even more at risk with a cluttered work area? Be sure that all of the relevant stick welding tools — your chipping hammer, wire brush, and electrodes, at the very least — are nearby but set up safely. (You can dig into all the details on stick welding equipment and tools in Chapter 5.)

You also need to take special care to ensure that your work area isn't in a place where other people may be working on non-welding projects or just passing by. You have a welding helmet on anytime you have a welding arc, of course, so your eyes are protected from the dangerous ultraviolet rays that are emitted by the arc, but other people nearby who aren't wearing protective eyewear risk vision damage if they look directly at the arc. Head to Chapter 3 for more on eye injuries caused by welding.

If the project you're undertaking does require that you stick weld outside of your shop in an area near other people, be sure that you use appropriate screening. You can buy welding screening material (usually made of vinyl) or use canvas covered in flat black or flat gray paint.

Eye injuries aren't the only danger you need to worry about when you're stick welding. The risk of electrical shock is high for stick welding, so to help reduce the risk of injury from shock, don't stand in water when you're stick welding. Also, make sure your clothing and gloves are dry.

Understanding stick welding electrodes

One of the most essential steps for a successful stick welding endeavor is picking out and using the most appropriate electrode for the job. (An *electrode* is a coated metal wire that should match the composition of the material you're welding.) Electrodes come in a dizzying array of materials, types, and sizes, and different combinations of those features result in different operating characteristics. (I go over some stick welding electrode basics in Chapter 5.) Choose the wrong combination for your project, and you're soon looking at a subpar, unsuccessful (and possibly dangerous) weld at best.

The ideal electrode for any job provides stability, a smooth bead, fast deposit of the welded material, maximum strength, easy cleanup, and very little *spatter* (the little bits of waste metal that come off of the welding arc and make the weld rough to the touch). Generally speaking, you can use all but the last two inches of an electrode.

No matter what kind of electrodes you work with, be sure that you store them in clean, dry space at room temperature. Otherwise, you end up with ruined electrodes, and that can be a frustrating (and expensive!) problem.

Electrodes usually come in steel or cardboard boxes. When you're opening a box of electrodes, be sure you heed the "Open this end" instructions printed on the box. If you open the other end you can easily damage the electrode.

Getting to know the characteristics of stick welding electrodes

A number of characteristics vary from one stick welding electrode to another, so consider the following list when you're choosing an electrode for your stick weld.

- **Type of material:** You must match stick welding electrodes to the type of metal that you're going to weld — that is, cast-iron electrodes weld cast iron, aluminum electrodes weld aluminum, and so on. The five basic groups of electrode material are mild steel, high carbon steel, cast iron, *nonferrous* (iron-free), and special alloy. If you're working with a metal that doesn't fall very clearly in one of those categories, read the information on the electrode packaging to confirm that it's recommended for what you're doing. If you're buying the electrodes from a welding shop, you can always ask a salesperson for advice.
- **Strength:** The strength of a metal is measured in *tensile strength.* The tensile strength of mild steel is about 40,000 pounds; high carbon or alloyed steels can have a tensile strength of up to 160,000 pounds, which is an extremely strong metal. The welds you make when stick welding need to be stronger than the base metal you're working on, so be sure that the electrodes you select are stronger than the base metals you're working on.
- **Size:** Stick welding electrodes come in a wide range of sizes, including $\frac{1}{16}$, $\frac{5}{64}$, $\frac{3}{32}$, $\frac{1}{8}$, $\frac{3}{16}$, $\frac{7}{32}$, $\frac{1}{4}$, and $\frac{5}{16}$ inch. Most stick welding is done with a $\frac{3}{32}$- or $\frac{1}{8}$-inch electrode. The smaller electrodes are useful for stick welding with low amounts of electric current, and when you're welding in odd positions. Believe it or not, larger electrodes are cheaper than smaller ones!

  The rule for electrode sizes is that the electrode core wire should never be larger than the thickness of the metal you're planning to stick weld.
- **Welding position:** You can find electrodes made for flat welding only, for both flat and horizontal welding, or for all welding positions, so be sure your electrode matches the position you plan to use. Check out "Assume the Position: Stick Welding in All Positions" later in the chapter for details on possible positions and my suggestions for appropriate electrodes.
- **Presence of iron powder:** To increase the volume of molten metal that you deposit during a weld, some electrodes have up to 60 percent iron powder mixed into their *flux* (the outer coating that surrounds the electrode's metal wire). The heat generated during the welding process converts the iron powder to steel, which ramps up the amount of metal that you deposit into the welded joint. Iron powder helps to create a smooth weld bead with outstanding finished appearance. The downside is that the presence of iron powder in an electrode can make stick welding much harder to control when you're welding in an unusual or difficult position.
- **Soft arc designation:** Some electrodes are designed with what's called a *soft arc.* These electrodes are for thin metal and for filling gaps or pieces that don't fit well together.

Looking at electrode classifications

You can find dozens of different kinds of electrodes for many types of stick welding. Luckily, the American Welding Society uses number codes to convey the different characteristics of each kind of stick welding electrode. The details of their classification system go beyond the scope of this book, but if you want to understand all of the ins and outs, check out the handy information that the folks from Lincoln Electric provide on the topic at `www.lincolnelectric.com/knowledge/articles/content/awsclassification.asp`.

Even if you don't want to know all the nitty-gritty details of the stick welding electrode classification system, understanding a few of the most commonly used stick electrodes for welding steel (one of the most commonly welded materials) pays off. Here are the ones I suggest committing to memory.

- **E6012 and E6013:** These electrodes offer a soft arc (see the preceding section), and they produce a smooth bead with thick slag covering the finished weld. Use them for thinner metals and ill-fitting joints.
- **E6010 and E6011:** These options are versatile electrodes that you can use on dirty, rusty, and oily metal without compromising the strength of the finished weld. You can use the E6011 either AC or DC *polarity* (the direction electricity flows), but you can use the E6010 with direct current electrode positive polarity only. (More information on polarity later in this chapter.) The E6011 is an all-purpose electrode and produces very little slag on the finished weld.
- **E7016 and E7018:** Both of these electrodes have iron powder in the flux, and that results in very strong welds. However, the weld puddle can be very hard to control for beginners who try to use these electrodes. See the preceding section for more on iron powder in electrodes.

Some stick welding electrodes become unusable when they're exposed to moisture in the air because they're dried at high temperatures in a low moisture environment when they're manufactured. Some of the kinds that are most affected include 7016 electrodes,7018 electrodes, and all stainless steel electrodes. If you leave these kinds of electrodes outside — even if they're in unopened cardboard containers — they pick up moisture from the air, and their quality is compromised. The worst part is that you can't easily tell when an electrode has absorbed moisture. You usually know only when your finished welds turn out weak and porous. The best way to keep many of your electrodes completely dry is to keep them in a heated electrode storage oven, like the one you can see in Figure 6-1.

Never put 6010 or 6011 electrodes in your electrode oven. They have a water-based flux that will dry out and crack, making them unusable.

Figure 6-1: Electrode storage oven.

Setting the Machine

No matter what kind of stick welding project you decide to take on, the process for getting your stick welding machine up and running for the job is the same. Here's what you need to do.

1. **Inspect the machine's cables and make sure they're tight.**
2. **If you're using a DC stick welding machine, set the polarity according to the requirements of the electrode you're using.**

   I provide more details on how to select polarity in the following section.
3. **Set the amperage on the machine to match what is required by the electrode you're using.**

   Amperage is the amount of electricity that the machine creates. You can find more on amperage later in the section.

   When you're just starting out in stick welding, set the amperage right in the middle of the range suggested for an electrode. For example, if an electrode's packaging indicates you should use it on a welding machine set between 90 and 100 amps, set your amperage right at 95.

Soon after you start your weld, you should be able to tell whether you need to turn up or reduce the amperage (while staying within the range suggested for the electrode, of course). If the electrode is too hot, reduce your amperage; if the electrode is sticking, increase your amperage. You can tell your electrode is too hot if you're burning holes through your *parent metal* (the piece you're welding) or the metal seems to be melting too quickly.

Choosing polarity

When you're choosing polarity, you have three choices for stick welding machines. Here's what each choice offers.

- **Alternating current (AC):** When you use *alternating current,* the electricity changes directions 120 times every second, which causes the welding current to flow from the electrode and the work evenly. That means that the heat produced by the current is also spread evenly (50 percent in the work and 50 percent in the electrode), and you get a weld bead with a perfect balance of *build-up* (how high your welding bead is off the surface of the parent material) and *penetration* (how deep into the joint the molten metal goes).
- **Direct current electrode negative (DC– or DCEN):** With DC–, the electrode is negatively charged and the work is positively charged. Two-thirds of the heat is in the work, and the other one-third is in the electrode. The only time I recommend using DC– is when you're going to be welding a thick plate in the flat position (more on welding positions later in this chapter), because DC– consumes your electrodes quickly.
- **Direct current electrode positive (DC+ or DCEP):** I think DC+ is the best choice when it comes to polarity. With DC+, your electrode is positively charged and the material you're working on is negatively charged, which really helps to make the most of your electrodes because when you control where the heat goes, you can deposit the weld exactly where it needs to be. When you use DC+, two-thirds of the heat is in the electrode and one-third of the heat is in the work. You should definitely try to use DC+ whenever your work allows.

Setting the amperage

Setting the amperage for a stick welding project is a very important step. Because it deals with the amount of electricity the machine creates, it has a direct relationship with the amount of heat you're working with. Higher amperage means more electricity and more heat.

By far, the most critical detail to pay attention to when you're setting the amperage is the operating range suggested for the electrode you're using. As I note earlier in the chapter, the packaging or label for all electrodes includes an amperage operating range, and you should take that suggestion very seriously. The ranges are set by the electrode manufacturers after years of performance and hundreds of hours of testing.

You can always adjust the amperage on your machine after you start a welding project, so don't feel like you're locked in to a specific amperage. There aren't any hard-and-fast rules that determine which amperage setting is best for you on any given stick welding project. Figuring out the amperage sweet spot is just a matter of tweaking the setting until you get it right, and that comes with experience.

Preparing to weld

You can't expect to do well with a stick welding project unless your welding machine is set up correctly and ready for action. Before you get started, put gloves on and clamp the bare end of the electrode into the *electrode holder* (also called the *stinger*), which is plugged into your welding machine. Fasten the ground cable to the piece of metal you're working on. Always keep your electrode holder dry.

Don't touch your welding table with an electrode in the electrode holder. That will cause a flash that can damage the surface of your welding table and also cause flash burns to your eyes if they're not properly protected as I explain in Chapter 3.

Whenever possible, drape or hang the electrode holder cable over your shoulder. That helps lessen the fatigue on your body, which allows you to weld more comfortably for longer periods.

Don't wrap or tie the electrode holder to you or any other piece of equipment. If someone walking or a piece of machinery driving by hooks your cable, you're going to go with it, and you can be injured. Also, wrapping the holder around something is dangerous because the DC current running through it will magnetize any iron-containing metal it's wrapped around.

If you take a break during a stick welding project (or when you finish the project), be sure to remove the electrode from the electrode holder and hang it (the electrode holder) on a hook on your welding table. Doing so will keep you from getting flashed or striking an arc inadvertently when you turn the machine back on.

You can see some common examples of electrodes in Figure 6-2. The *covered electrode* is a type of filler rod used in arc welding. You can identify electrodes by the numbers printed on the *flux,* which is a covering surrounding the core wire that determines the composition of the electrode. The original purpose of coating the wire was to shield the arc from oxygen present in the air. After that, different materials were added to the flux to help stabilize the arc and improve the strength of the weld. Iron powder is also added to flux so you can deposit more weld in less time.

The flux on electrodes for welding mild steel have up to 12 ingredients. These include cellulose, metal carbonates, titanium dioxide, ferromanganese, ferrosilicon, clay, gum, calcium fluoride, mineral silicates, iron, iron powder, and alloying metals such as nickel, chromium, molybdenum, and manganese.

Figure 6-2: A layout of several electrodes.

Striking and Maintaining an Arc

When your work area is set up and you have your electrodes in order, it's time for you to get started on your stick weld. The very first step in stick welding is *striking an arc* to get the electricity to jump *(arc)* from your electrode to the *parent metal* (the metal you're going to be welding). Striking an arc is an extremely important skill. It's just as important to a welder as swinging a hammer correctly is to a carpenter. The idea is that you strike the arc at the starting point of the weld and maintain the arc for the length of the weld, thereby fusing the materials together. The molten metal from the electrode is deposited in the joint, and the intense heat means that some of the parent metal melts and mixes with molten electrode to produce a sound weld.

Before you strike any welding arc, be sure you're wearing your welding helmet so your eyes are protected!

But before I get into the details of how to strike an arc, I need to fill you in on five factors that can greatly influence the success of any stick welding project.

- **Current:** The type of current or polarity you use is either AC or DC. If you use the wrong current, the weld doesn't deposit correctly. Flip to "Choosing polarity" earlier in the chapter for more on polarity considerations.
- **Length of arc:** The arc of electricity that you maintain between your electrode and the parent metal should be about the length of (and no longer than) the diameter of your electrode's core wire. A proper arc makes a continuous frying sound. Welding with a long arc makes an uneven crackling sound, and sometimes the arc can go out. It creates excessive weld spatter, and your weld bead is uneven and wide. If your arc is too short, it makes a buzzing sound. Your electrode sticks to the parent metal, and your weld bead is too narrow. You want to end up with a finished weld that's no wider than 2.5 times the diameter of your electrode's core wire.

  If your electrode sticks to the parent metal, break it loose by bending the electrode holder slightly or giving it a quick twist. If that doesn't work, release it from the electrode holder or shut off the welding machine. Then break it off from the work, reinsert it into the electrode holder, and try again.
- **Angle of electrode:** The angle created by your electrode and the parent metal should be no more than 5 to 15 degrees in the direction you're welding. (That doesn't hold true for vertical welding, and I get to that a little later in this chapter.) If you don't get the angle right, your welding penetration won't be correct.
- **Manipulation of the puddle:** The puddle of molten metal that you create as you weld can have gas pockets in it that compromise the strength of your weld. You can eliminate the gas pockets by moving the puddle with a slight (as in extremely subtle) side-to-side motion. Don't overmanipulate the puddle — that makes your weld bead too large.

- **Speed of travel:** The speed at which you move the electrode across the parent metal is critical. If the travel speed is too slow, you end up with too large a weld. You waste your electrodes, and the excessive heat causes warping. If you're trying to rush and the travel speed is too fast, you end up with a weld bead that's far too narrow.

To help you remember these five factors, think about the word *clams* (*c*urrent, *l*ength of arc, *a*ngle of electrode, *m*anipulation of the puddle, and *s*peed of travel.)

While keeping in mind the factors, you can go about actually striking an arc. You can use one of two methods to strike the arc: tapping and scratching.

- To use the *tapping* method, move the electrode straight down, tap it on the parent metal, and then draw it back.
- To use the *scratching* method, you simply scratch the electrode on the surface of the parent metal as though you're striking a wooden match. Whether you use the tapping or scratching method, as soon as you have the arc established, you need to immediately move the electrode down to a distance from the parent metal that's equal to the diameter of your electrode's core wire.

When you're just beginning to stick weld, feel free to practice striking an arc over and over by striking the arc and then welding a small circle about the size of a dime. Then break the arc and try again. A master welder can strike an arc blindfolded, with one hand, upside down, in a snowstorm. Okay, that may be a bit of a stretch, but rest assured that any practice you put into figuring out how to quickly and effectively strike an arc won't be wasted time.

Assume the Position: Stick Welding in All Positions

Not everything you weld comes set up in the most comfortable position. Sometimes you may be forced to weld in a challenging position. In addition to welding on a flat surface (like your welding table), you may be faced with welding horizontally, vertically, or even overhead.

No matter what position you stick weld in, your goal should be a strong, clean finished weld. To get an idea of what good and bad welds look like, take a look at the welding piece of metal in Figure 6-3. From top to bottom, the examples are: good; too fast; too slow; too short of an arc; too long of an arc; too hot; and too cold.

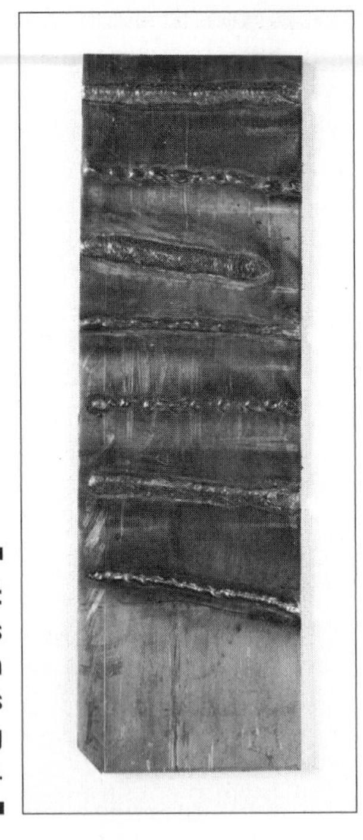

Figure 6-3: Examples of seven stick welds of varying quality.

Welding on a flat surface

Stick welding on a flat surface is the easiest position to weld in (as is the case with all welding processes), and it's a great way to practice stick welding. Just follow these steps.

1. **Put a 6-inch square piece of metal ¼ or ⅜ inch thick on your welding table.**
2. **Insert a ⅛-inch E6011 electrode into your electrode holder; set your welding machine with the current and amperage specified on the electrode packaging or label.**

   The amperage should be about 90 amps. You're now ready to run a bead.
3. **With a soapstone, draw a series of lines on the parent metal approximately ½ inch apart.**

   Soapstone is a special material that allows you to write on metal, which you can't do with regular pens or pencils.

4. **Holding the electrode in a vertical position and slanting it slightly in the direction you're welding, strike the arc and move the electrode across the parent metal in a continuous line that goes over each line you've drawn.**

   See "Striking and Maintaining an Arc" earlier in the chapter for more on firing up your arc. As you move the electrode, go from left to right if you're right-handed, or right to left if you're left-handed — welding across your body is easier than welding toward or away from your body.

5. **Move the electrode side-to-side rapidly enough that the deposited metal penetrates into the base metal.**

   As I note earlier in the chapter, if the welding machine is set properly and you maintain the arc ⅛ inch or less, you hear a continuous frying noise. That's a good sound, so remember it!

6. **To fill the crater at the end of the bead, draw the electrode up slowly and backward over the finished weld.**

   Slowly drawing up the electrode allows the crater to fill with metal, and moving back over the weld puts the crater on top of the bead. If the crater starts to burn through the parent metal, stop welding, chip out the crater, and restart the arc. You can fill the crater by restarting as many times as necessary.

To demonstrate how important correctly setting the amperage is, turn it down five amps and weld another bead just as I describe in this section (after you've mastered that process and you're acquainted with your machine). Do you notice any differences when you run the new bead? Dial down your amperage by five more amps and try again. Does the electrode stick all the time? This hands-on experiment helps you experience just how vital the correct amperage is.

The appearance of the finished weld indicates how good the weld is. If the finished weld is clean and clear of any slag and the edges are full, it's a good weld.

Going vertical

Stick welding in a vertical position is trickier for most people than welding on a flat surface (see the preceding section), but if you keep in mind some basic steps and get in some practice time, you should be able to pick up the skill. It's a useful position because it allows you to weld on equipment or structural metals that you can't lay flat on their sides. There are two types of vertical position stick welding: vertical-up and vertical-down.

- **Vertical-up:** You start at the bottom of a joint and work your way to the top; when you're welding in the vertical-up position, your electrode should be pointed at an upward angle of 5 to 15 degrees. You use vertical-up on materials thicker than ⅛ inch.
- **Vertical-down:** You start at the top of a joint and work your way to the bottom at an angle of 65 degrees. You use this method for materials that are ⅛ inch thick or less.

I explain the basics of both methods in the following sections.

Vertical-up stick welding

To practice stick welding in the vertical-up position, follow these steps.

1. **Get a ⅛-inch E6011 electrode and a 40-inch-long piece of angle iron that's 3⁄16 inch thick, with both sides of the angle at 2 inches wide.**
2. **Use the technique for welding in a flat position that I outline earlier in this chapter to weld the angle iron to the corner of your welding table, like the example in Figure 6-4.**

   Now you have a piece of angle iron going from the top of your table straight up 40 inches.

Figure 6-4: Welding an angle iron to your welding table.

3. **Clamp a plate of metal that's at least 6 inches square and ⅜ inch thick to your angle iron so it's in the vertical position at eye level.**

   If you don't have a clamp, you can tack weld the plate in place. *Tack welding* simply involves making a small (less than 1 inch) weld to hold something in place temporarily. Grab the plate with one hand and pull on it to make sure it can stay put. If it's loose, it may come off during the weld if you have to remove a stuck electrode.

4. **Strike an arc and do a half dozen dime-sized welds in the vertical-up position on the side of the plate.**

   Check out Figure 6-5 to see an example.

Figure 6-5: Practicing vertical-up welding.

When you're comfortable with that technique, use your soapstone to draw four or more lines up from the bottom of your plate. Strike an arc at the bottom of the plate, maintain the arc, and slowly weld from the bottom to the top with a slight side-to-side motion, pausing only on the sides. Your 14-inch-long ⅛-inch-diameter electrode will deposit 5½ to 6 inches of welded material. Remember, the bead should be no wider than 5⁄16 inch. Repeat this process until you're comfortable with it and the bead looks good.

Vertical-down stick welding

Electrode selection is important when you're vertical-down stick welding. You need to account for the thickness of the metal you're welding, and I recommend practicing with an E6013 electrode.

The technique for vertical-down stick welding starts with the angle of your electrode.

1. **To start the weld, tilt your electrode 65 degrees upward, starting at the top of the joint you're planning to weld.**

   Vertical-down stick welding is faster than vertical-up stick welding because gravity is more on your side.

2. **Strike your arc and keep it short.**

3. **Weld downward and be sure to form the bead fast enough to keep the molten slag and metal from running ahead of the *crater* (the ending point of the weld).**

   That helps to keep the weld nice and clean. Refer to Figure 6-6 for a good vertical-down stick welding example.

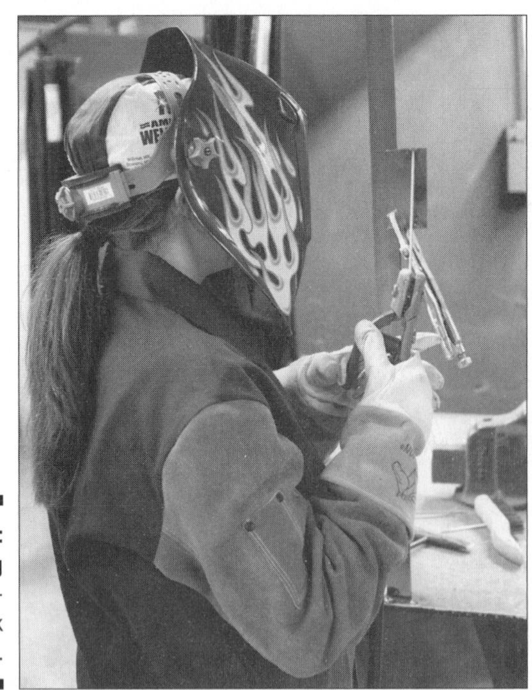

Figure 6-6: Practicing vertical-down stick welding.

Exploring horizontal welding

You'd be surprised how infrequently you're able to weld on a flat surface (like a welding table) after you start getting into a wide variety of stick welding projects. Most welders really like to weld on a flat surface, but they also know that the horizontal position is much more common. In many cases the metal you need to weld can't be manipulated into the flat position. A weld in the *horizontal position* occurs when the weld runs on a line parallel with the horizon.

If the difference between welding on a flat surface and welding in the horizontal position isn't clear, think of the former like writing on a desk and the latter like writing on a chalkboard.

One key difference between welding on a flat surface and welding in the horizontal position is that you need to maintain a shorter arc and lower amperage (about 5 to 10 amps lower) when you're welding in the horizontal position.

Here's how you can get started welding in the horizontal position.

1. **Clamp a 6-inch-square plate of metal ⅜ inch thick to your welding table so the broad side of the plate is facing you and draw horizontal lines on the plate about ½ inch apart with your soapstone.**
2. **Using an E6011 electrode, strike the arc and maintain a 5- to 15-degree angle with your electrode in the direction of travel as you start welding across the plate, following the lines.**

   If you're right-handed, weld left to right across your body just like you're reading. If you're left-handed, weld in the opposite direction. Be certain to maintain the correct arc length across the surface of the metal. Keeping a nice tight arc helps keep the molten puddle from sagging (and a sagging puddle won't fill the top edge of the weld). Using a narrow weaving motion and pausing at the top as you weld reduces the chances that your weld sags.

   If you come to the end of the joint and you still have some of your electrode left over, don't throw it away! You can always use it for more practice.

I show you what it looks like to weld horizontally in Figure 6-7.

Figure 6-7: Welding horizontal.

Reaching overhead

Stick welding in the overhead position is the hardest position to master. It's usually more physically taxing than the other positions; it's also difficult because you're working against gravity, and that can pull your weld down. When you weld overhead, your puddle has a tendency to droop, making it hard to get good penetration. However, despite these challenges, you can rest assured that getting good at stick welding overhead involves the same key ingredient that you need to master the other positions throughout this chapter: practice. And you'll be glad you put in the practice when you're forced to weld in the overhead position. For example, if you need to weld the bumper hitch on your car, you can't flip over the car to make a flat weld. Here's how to get started.

1. **With an E6010 electrode, weld a piece of angle iron to your welding table like I explain in "Vertical-up stick welding" earlier in the chapter.**

   The angle iron needs to reach up more than one foot above your head.

2. **Tack weld a ⅜-inch-thick, 6-inch-square plate of metal to the angle iron up above your head and then practice welding two or three dime-sized beads.**

   You can read more on tack welding in the earlier section "Vertical-up stick welding." Be sure your electrode is sticking straight out of the electrode holder. Hold the electrode holder with your knuckles up and palms down to prevent spatter from lodging around your hand. To get a visual, check out Figure 6-8.

   Overhead welding can be hard work, and it requires a pretty awkward stance. To help you maintain some level of comfort, I suggest placing the electrode holder cable over your shoulder. That lessens the weight you're holding up with your arms, which should allow you to weld for longer periods of time without resting.

3. **Using your soapstone, reach up and draw three or four lines horizontally across the surface of the plate.**

4. **Keeping your feet a little wider than shoulder width and your arms close to your body, reach up and weld a bead on the lines you drew on the plate.**

Figure 6-8: Practicing overhead stick welding.

When you're welding overhead, check your clothes often to make sure falling molten metal hasn't stuck in any cracks or folds. It's also a good idea to wear a welder's hat and heavy shoes. (Read more about choosing the right clothing and safety gear for welding in Chapter 3.)

Chapter 7

To Tig or Not To Tig: Understanding Tig Welding

In This Chapter

- Introducing tig welding elements
- Exploring some pros and cons of tig welding
- Making sense of tig welding equipment

If you're looking for an alternative to stick welding (see Chapters 5 and 6) that's long on versatility but short on mess, look no farther than tig welding (also called *gas tungsten arc welding* or GTAW.) Tig is a more complex welding process, and it's quite a bit more expensive than stick welding, but if you can get the hang of it and don't mind spending the extra money, the results can be outstanding.

In this chapter I mention some of the differences between stick welding and tig welding, but don't let that make you lose sight of the fact that tig welding is still an arc welding process and that the basic principles used to melt the metal and make the weld are the same.

Tig welding is the most versatile welding process; you can use tig to join metals of almost all types — from steel to brass to aluminum and just about everything in between. Tig is also great for welding materials in a huge range of thicknesses. The basic idea behind tig welding is the same as stick welding: You use an arc of electricity to heat up metal to the point where it melts into a puddle, and you manipulate that puddle to join metals together. In this chapter, I provide a rundown of the tig basics that you need to know to really wrap your brain around the process, the specific equipment you need to get started, and the reasons why you'd choose to use tig rather than another welding process like stick or mig (which I cover in Chapters 9 and 10).

Tig welding is a lot easier to master if you've practiced oxyfuel welding in the past. For more information on oxyfuel welding and other special welding processes, check out the chapters in Part IV.

Taking a Closer Look at Tig Welding Components

Put simply, *tig welding* involves using a non-consumable tungsten electrode — *non-consumable* means that the electrode doesn't get melted during the weld — to create an arc for welding, and protecting the welding area from contaminants with a shielding gas. A *shielding gas* doesn't react with metals, and you apply it to the welding area in a constant flow while you're welding. (I go into more details about shielding gases later in this chapter and in Chapter 8.)

The tungsten electrodes don't provide the filler metal (as the electrode does in stick welding). The filler metals used in tig welding come in the form of a separate rod.

Tungsten is used for tig welding electrodes because it has an extremely high melting point; you have to have temperatures above 6,000 degrees Fahrenheit to melt it! (That's why tungsten is also used for lightbulb filaments.) So tungsten electrodes can easily heat other metals to their melting points without the electrodes themselves actually melting.

The other important aspect of tig welding is the shielding gas. These gases — argon and helium are the most common choices — don't react with molten metals. When you apply them to the welding area in a continuous flow, they protect the weld from coming in contact with the air, which contains a number of gases that *do* react with molten metals in all kinds of annoying ways that can compromise the strength and appearance of your work.

Considering the Advantages and Disadvantages of Tig Welding

If you ask me, the biggest advantage that tig welding affords is the quality of work that you can achieve across almost all types of metal. With the exception of a handful of metals that have very low melting points, you can join all the metals that you'd ever want to weld in an effective and attractive way by using tig welding. It's remarkably dynamic! But the advantages of tig welding don't stop there. Following are some other pros of the tig welding process:

- **You can weld metals of many different thicknesses.** That includes very thin materials that can be extremely difficult — or even impossible — to weld using other welding processes. You can tig weld using an amperage range of as low as 5 amps to more than 800 amps. (That's a huge range!)

- **It produces joints that are ductile, stronger, and more resistant to corrosion than the joints you can achieve with other types of welding.** *Ductile* joints are less brittle and less likely to break.
- **The tig welding process is extremely clean.** Tig welding produces very little spatter, smoke, sparks, or fumes and is far less messy than what you can expect with other welding techniques.

Even though tig welding produces less smoke and fewer fumes than stick welding, you still need to ensure proper ventilation, especially when tig welding copper alloys (which I discuss later in the chapter). The fumes created when you tig weld copper alloys are extremely toxic.

- **The visibility you can enjoy while tig welding is second to none (well, at least no other welding technique).** Because tig welding produces very little smoke and uses colorless shielding gases, you can see your work while tig welding in a way that just isn't possible on a comparable stick welding job.
- **Tig welding requires very little finishing.** You don't spend your time using the chipping hammer and wire brush to remove slag in the way you typically do when stick welding.
- **You can tig weld in any position.** I discuss tig welding positions more in Chapter 8.

After reading a list of advantages like that, you may be inclined to just go ahead and fall head over heels in love with tig welding. Not so fast! As with every other welding process, tig welding has some definite drawbacks. Here are the major disadvantages you can expect from tig welding.

- **Tig welding is slow.** It goes faster after you really get good at it, but it's still slow when compared to most other welding processes.
- **You don't get good at tig welding overnight.** It's quite a complicated process that takes most people lots and lots of practice before they really see success.
- **The equipment and supplies needed for tig welding are expensive.** If you plan on getting into tig welding, you can expect to pay quite a bit more than what you'd pay to get a stick welding operation up and running. See "Stocking the Shop: Examining Tig Welding Equipment" later in the chapter for more on tig supplies.
- **Tig welding can be tough to do outside of a controlled environment like a welding shop.** Because you have to keep a constant flow of shielding gas moving over your weld area, tig welding outside — especially if it's windy — can be a real challenge.

Brushing Up on Tig Welding Basics

You can't take full advantage of the versatility of tig welding if you don't have a good feel for the metals you can weld with it and you don't make an effort to control your tig welding environment. In the following sections, I provide you with some metal know-how and offer a few pointers to help you ensure that your surroundings don't get in the way of a successful tig weld.

Showing your metal: Looking at a few metals for tig welding

You can use tig welding for all kinds of metals, but here are a few of the most popular ones, along with some suggestions on how to tig weld them. You can find more information on the welding machine settings mentioned in "Controlling current and amperage" later in the chapter.

- **Aluminum** is one of the most common metals people use to tig weld. It has a very low melting point and conducts heat extremely well.

  When you tig weld aluminum, I recommend that you set your tig welding machine for AC output with continuous high frequency. Using high frequency allows you to start welding without touching the tungsten to the piece you're welding, and that means your tungsten doesn't get contaminated. Check out the later section "Controlling current and amperage" for more on electrical current classifications.

- **Magnesium** is extremely similar to aluminum in terms of the way you treat it when tig welding — both metals have a low melting point and conduct heat extremely well. The big difference is that when you weld magnesium, the weld area is weaker than the surrounding metal as a result of the heat from the welding process.

- **Copper alloys** include brass and bronze, as well as mixtures of copper with nickel, aluminum, and silicon. You can tig weld all these copper alloys by using a direct current electrode negative (DC–; see Chapter 6) setting on your welding machine.

- **Stainless steels** are prime targets for tig welding. When you're working with stainless steel, you typically achieve a better result if you keep your arc nice and short. Be sure to use a filler rod that has a higher chrome content than the metal you're welding.

  If you have problems with *cracking* (cracks caused by expansion or contraction that takes place during the welding process) on your stainless steel tig welds, try preheating the metal to 400 degrees Fahrenheit before you begin to weld next time. That should help prevent some of the cracking.

- **Mild steels** come in many different alloys and types, and you can easily weld them with the tig welding process. Mild steel filler rods must have *deoxidizers* (substances added to keep them from breaking down) added to them to ensure a high-quality tig weld, and you should use DC– current with a high frequency start. You should also use a tungsten electrode that's 2 percent thoriated. (More on that at the end of this chapter.)

Some metals need cleaning before you can tig weld with them. For example, you have to mechanically clean steel to make it suitable for welding. Aluminum can be very sneaky because it sometimes appears completely clean when it's actually covered (or partly covered) in a coating of aluminum oxide, which forms when aluminum is exposed to the air, that you have to scrub off with a wire brush. Aluminum oxide has a much higher melting point than pure aluminum, so when you start welding, the aluminum underneath the coating melts before the coating melts; that can end up causing some messy (and even dangerous) situations.

Taking steps to ensure quality welds

Of all the arc welding processes, tig welding yields the highest-quality finished welds. That's why tig welding is commonly used in industries like automobile customization and aerospace. But you can't expect to just fire up your tig welder, strike an arc, and magically end up with beautiful welds. You need to take several steps to ensure that you're giving yourself the best possible chance to make a great tig weld. Luckily, the following list contains exactly those steps.

- **The best quality weld is the one that's accomplished in a safe environment, without injury or unnecessary risk.** Flip to Chapter 3 to read up on the safety information that you need to keep in mind when you're getting set up for tig welding (or any other kind of welding).
- **Keep your work space clean!** You don't end up with strong, good-looking tig welds if the welding area is dirty. In particular, your work table, filler rods, and gloves should be completely clean before you get started on a tig welding project.
- **Consider your shielding gas.** Whether you end up using argon or helium as your shielding gas, you need to make certain that your gas supply is high quality and low moisture. When you buy a tank of shielding gas at your local welding supply store, confirm with them that the gas you select has received high marks from other welders and doesn't have a reputation for moisture issues. (Moisture in the air can cause some shielding gases to move away from the molten pool.)
- **Avoid *porosity* (tiny holes) in your finished welds by ensuring that the metal you're welding is dry, using clean filler rods, and keeping the shielding gas flowing at all times while welding.**

If you make a commitment to yourself to take the steps to ensure a good tig weld every time you get started on a new project, you give yourself a great chance at achieving success. With a little practice, you'll be making beautiful finished tig welds in no time.

Stocking the Shop: Examining Tig Welding Equipment

Tig welding equipment is important to any good tig operation. As with any welding style, you need reliable equipment and supplies if you want to achieve the best results, so you should be familiar with how to recognize (and use!) good tig gear. For a quick look at what tig equipment looks like, check out Figure 7-1.

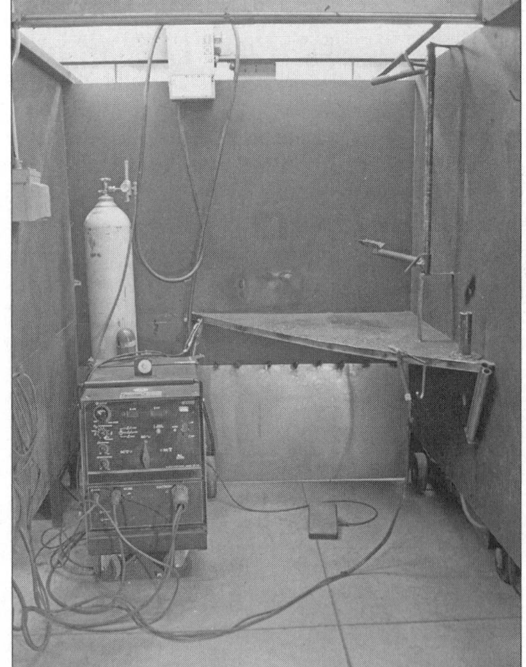

Figure 7-1: A tig welding setup.

The items you need for tig welding are expensive, and there aren't many ways around that fact. You can spend two or three times as much on tig welding machines, for example, as you can on a stick welding machine. You can always take steps to get less-expensive equipment — buying used or trading in another machine to get a reduced price, for example — but you're still going to end up thinning out your wallet to gear up for tig welding.

Another thing to keep in mind when it comes to tig welding equipment: A lot of variables can affect your tig welds in a lot of different ways. Some of the most important variables include tungsten type and size, shielding gas type and nozzle size, torch angles, filler rod type, current, and correct clamping. If some of those considerations don't make a whole lot of sense to you, fear not. I go into more detail in the following sections and throughout much of Chapter 8.

The manufacturer's recommendations for tig welding equipment make for excellent guidelines, and you should stick to them whenever possible to help control those variables. Check the instruction manuals and specification sheets on your equipment and supplies to make sure you have all your settings where they need to be in order to help ensure a quality weld. Pay special attention to shielding gas flow rate, *polarity* (the direction in which the electricity flows), and amperage.

Considering fully equipped tig machines

Compared to other welding machines — especially stick machines — tig machines are pretty complex and have a lot of controls and other bells and whistles. Tig welding machines are also classified as *constant current* machines, which means they can produce a constant stream of electricity for welding without changing the voltage much at all. That allows you to make perfect welds on very thin materials.

Tig welders are complicated, dynamic machines, and when you buy one you pay for that versatility. However, if you're planning to do both stick and tig welding in your welding shop, you can kill two birds with one stone (and save some money) by buying only a tig welding machine. Believe it or not, you can use tig welding machines for stick welding.

The following list shows you some of the features and controls you should be familiar with as you shop for or use a tig welding machine.

- A control (usually a switch) that allows you to specify whether you want to use the machine for tig or stick welding.
- A current control that allows you to control the machine's current by using a dial on the face of the machine.
- Three starting modes:
    - **High frequency start only** is used only for welding ferrous (iron-containing) metals such as steel or stainless steel.
    - **High frequency continuous** is used for aluminum and magnesium with AC current.
    - **High frequency off** is used when you're stick welding with your tig welding machine.

- An amperage control dial.
- A selector that allows you to choose what kind of electrical output you want to use (AC, DC+, or DC–).
- Timers that control when shielding gas flow starts before you strike an arc to begin welding and when flow stops after you're no longer maintaining an arc. These are usually called *pre-flow timers* and *post-flow timers*.
- A cleaning mode, which is really helpful when you're welding aluminum and you want to get maximum cleaning.

Thinking about tig torches

Tig torches are a critical part of any tig welding setup. A tig torch does all the following things:

- **Holds the tungsten electrode:** Tig torches allow you to use tungsten electrodes of varying diameters. The torches have a *collet,* which is a piece that holds the electrode. You use different size collets for electrodes of different diameters.
- **Provides an electrical connection to the tungsten electrode:** You can control the current with a switch on the torch or a foot pedal.
- **Provides shielding gas flow to protect the welding area:** Shielding gas is fed through a nozzle to the welding area. The gas comes out of a ceramic cup that's fitted to the torch; you can use different ceramic cup sizes to produce a range of shielding gas flow rates.
- **Protects you by insulating the dangerous electrical connections:** Pretty important considering the massive amounts of electricity used in tig welding.

Tig torches are exposed to quite a lot of heat, so they're usually either air or water cooled. You can get a good look at a tig torch example in Figure 7-2.

Selecting and managing shielding gas

The shielding gas used in tig welding comes in pressurized cylinders that you can buy at your local welding supply store. The cylinders are usually pressurized to 2,200 pounds per square inch (psi).

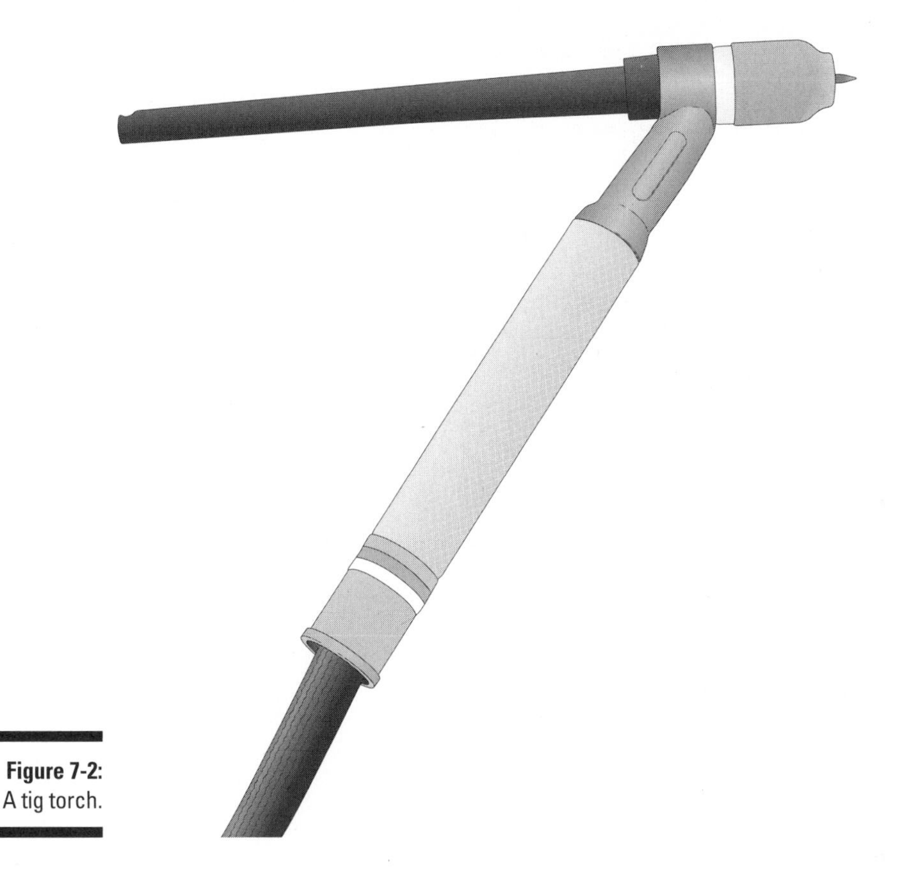

Figure 7-2: A tig torch.

The gas is controlled by special equipment — called *flow meters* — that regulate the flow gas around the arc during welding. Flow meters are usually calibrated to allow you to adjust the shielding gas in cubic feet per hour (cfh). The amount of shielding gas flow you need for a particular project depends on the metals you're welding; read through your tig welding machine manual for more details.

The flow rate of your shielding gas should be high enough to protect the tungsten electrode and the molten pool while you're welding. If you use too much gas, you can get porosity in your tig welds. One good way to judge whether you're using the right level of shielding gas flow is to examine your tungsten electrode after welding. If it's smooth and rounded on the tip, you've been using the right shielding gas flow. However, if the electrode is burned or deformed, you're not using enough shielding gas.

Wind can be a very serious problem when you're tig welding! If the area you plan to weld in has wind (or any air movement, for that matter) above two miles per hour, you need something to block the wind so that it doesn't interfere with your shielding gas flow. You can buy a windscreen at your welding supply store or online.

Figure 7-3 shows you a tig welding machine.

Figure 7-3: Tig welding machine.

As I note earlier in the chapter, the most common shielding gases are argon and helium. Argon is the overwhelming favorite, and you should use it for the vast majority of your tig welding endeavors. Why is argon so popular? For starters, unlike helium, argon is heavier than air (which is why you never see anyone filling up birthday-party balloons with argon gas). This fact is significant because it allows the gas to shield your welds better than helium does. Following are a few more reasons why argon is the preferred shielding gas for most tig welders (including yours truly):

- You can more easily avoid burning through thin materials.
- It gives you better control than helium does when you find yourself in an unusual position.

- It produces a better cleaning action when you're welding aluminum with AC current.
- You can use a higher voltage while tig welding when you use helium as your shielding gas.

With all these excellent qualities of argon, you may be wondering when you'd ever want to use helium as your shielding gas. Well, even though it's more expensive, helium allows you to use a higher voltage setting on your tig welding machine while you're welding. That means you can make welds at a higher speed.

If you're short on time and you need to be able to make tig welds quickly, using helium as your shielding gas helps you weld more quickly than argon can.

Controlling current and amperage

Most tig welding machines allow you to choose whether you want to use *alternating current* (AC), which flows back and forth at regular intervals in a circuit, or *direct current* (DC), which flows in one direction. You can simply flip a switch to toggle back and forth between those two options. The most important factor in determining whether you should use AC or DC is the kind of metal you're planning on tig welding — some metals can be joined only with AC, and others require DC. For example, use AC (continuous high frequency) when you're tig welding aluminum and magnesium. When you're working with stainless steel, brass, cast iron, mild and high carbon steel, and copper, be sure to use DC–. The only time that you should really use DC+ is when you're welding extremely thin materials; be sure to read your tig machine's instruction manual for more details.

Most tig welds are created with DC– or AC because the tungsten electrode is negative and the metal you're welding is positive. In those cases, most of the heat — usually about 70 percent — is in the piece that you're welding, and that's exactly where you need it to be.

Tig welding machines allow you to start an arc with a switch or button, and these are typically located in a foot or hand control such as the ones in Figure 7-4. The foot control also regulates the amount of amperage delivered to the tig torch. (You can think of it like the gas pedal on your car; the farther down you push it, the more power you get.) A hand control is a wheel mounted on the side of the machine. You can turn that wheel with your thumb or forefinger while welding to increase or decrease the amount of amperage going to the tig torch.

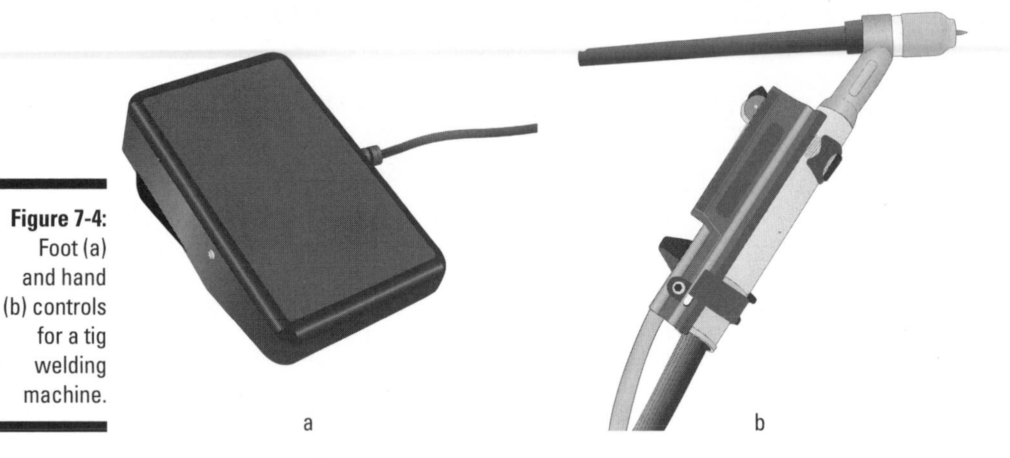

Figure 7-4: Foot (a) and hand (b) controls for a tig welding machine.

Selecting filler metal

Some tig welding projects (called *autogenous welds*) don't require a filler metal because it's possible to use only the metal already present to complete a strong joint. As a general rule of thumb, you can tig weld without a filler metal if you can maintain a pool of molten metal without melting through the piece of material you're welding on, which depends mostly on the speed at which you're welding. If you're welding a bead too slowly, you'll likely melt through the material and have too wide a weld bead. However, if you overcorrect and weld too quickly, you'll end up with a narrow (and probably weak) bead.

If you figure out that you need a filler metal, that material comes in the form of a *filler rod,* which is simply a long, thin, cylindrical rod. Filler rods come in many different sizes and kinds of metal — Figure 7-5 illustrates a few examples. Special tig welding rods are available; these rods contain more deoxidizers, which yield sounder weld joints.

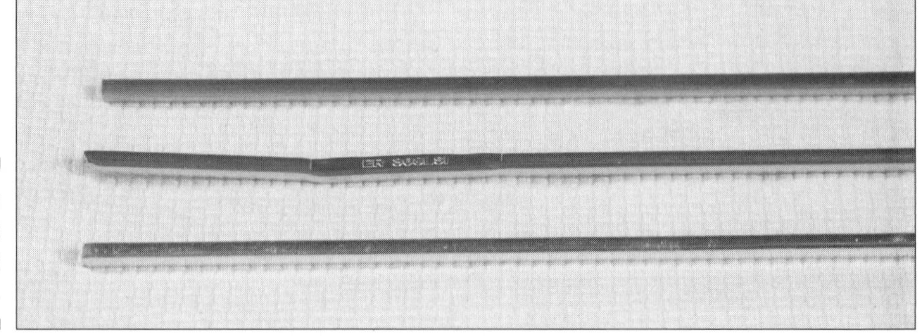

Figure 7-5: Filler rods for tig welding.

Filler metals have to be the same composition as the *parent metal* (metal your project is composed of) that you want to tig weld. Steel rods weld steel, aluminum rods weld aluminum, and so on. The most common filler rod used for tig welding steel is the ER70S-6. If you're going to be tig welding stainless steel, the best filler rod for you is probably the ER308. And if you're working with aluminum, check out the filler rods from ER4043 through ER5356. (Those classifications are made by the American Welding Society.)

You can't use copper-coated filler rods intended for oxyacetylene welding for tig welding. The copper coating contaminates the finished weld. (Flip to Chapter 13 for more information on oxyacetylene welding.)

Exploring tungsten electrodes

Although tungsten electrodes are non-consumable and don't melt and become part of your weld, the intense heat of the arc may cause a small amount of the tungsten electrode to be transferred across the arc. When that happens, some tungsten becomes included in the finished weld.

Tungsten electrodes come in diameters that range from 1/20 to 1/4 inch, but the most commonly used sizes are 1/16, 3/32, and 1/8 inch. I go into more detail on tungsten electrode selection in Chapter 8, but as a general rule, the amount of current that needs to flow through an electrode for a given tig project determines the size of the electrode necessary for it.

Some welds require you to shape the tip of the tungsten electrode. Tig welding using DC requires that you sharpen the tip to a point, but AC tig welds require a rounded tip.

Some tig welding electrodes are pure tungsten, and others are a tungsten and thorium alloy. Thorium is added to increase the amount of electrical current that the tungsten can carry, and it helps to keep the tungsten cooler during the welding process. A *thoriated electrode,* as it's called, stays cooler and holds up better during critical welds.

The two most common tungsten electrodes are green and red. Green electrodes are made of pure tungsten, and you use them for tig welding aluminum and magnesium. Red electrodes are 2 percent thoriated, and you use them for tig welding everything else (mild steel, stainless steel, and so on).

Chapter 8

Trying Out Tig Welding

In This Chapter

- Getting your tig welding supplies in order
- Looking at settings for various kinds of tig welding projects
- Starting the tig welding process by striking an arc and checking your grip
- Executing various tig welds

Tig welds are smooth, strong, and uniform. The tig welding process produces very little smoke, spatter, and *slag* (debris), so the welding environment is cleaner and often more comfortable for the welder. You can tig weld in all positions, and you can tig weld a huge range of metals in an array of thicknesses. Put simply, tig welding is terrific.

In Chapter 7, I offer a rundown of tig welding basics so you can get a feel for how the process works and what you need in order to do it successfully. In this chapter, I take the tig conversation a step farther and clue you in on how to actually make a tig weld. As with most of the other welding processes, a successful tig weld starts with quite a bit of preparation — you can't expect to produce a nice tig weld unless you're all set up to handle the job. I start this chapter with a close look at tig welding prep, and I finish it off with a breakdown of the steps you need to take to perform a tig weld.

Getting Your Welding Setup Tig-ether

Tig welding is a pretty complicated process, but you can make it much, much easier if you have your tig welding operation set up in the best possible way. That's what this section is all about. Read on to get the skinny on selecting tungsten electrodes for tig welding, making sure your shielding gas is all set up and ready to go, and figuring out the details of tig welding filler rods. (I go through quite a few of the details on tig welding machines in Chapter 7.) When you get ready to dive into a tig welding project, your setup needs to resemble what I discuss in Chapter 7.

Taking care of tungsten electrode details

Like stick welding, tig welding requires the use of electrodes. However, unlike stick welding, tig welding involves *non-consumable* electrodes made out of tungsten. (Tungsten is a metal with a very high melting point of 6,000 degrees Fahrenheit or more, so it doesn't melt and become part of your weld like stick welding electrodes do.) Tungsten electrodes transmit heat from the power source (your tig welding machine) to the work piece.

Tungsten electrodes for tig welding are either pure tungsten or tungsten mixed with another material like thorium, zirconium, or cerium. Those materials can make the tungsten more durable and able to carry larger amounts of electricity for more powerful welding.

To help you figure out which tungsten electrodes are right for your tig welding projects, Table 8-1 gives you some information on the types and applications of a few electrode types. As you take a look at the table, keep in mind that the codes in the "AWS Classification" column come from the American Welding Society. These codes can help you to make sure you're getting exactly what you need when you hit your welding supply store. (Check out "Choosing an electrical current and striking the arc" later in the chapter for more on the types of polarity listed here.)

Table 8-1 Tungsten Electrodes Types and Applications

Type of Electrode	*AWS Classification*	*Color*	*Polarity*	*Applications*
Pure	EWP	Green	AC	Aluminum, copper, and alloys of magnesium
1% Thoriated	EWTh-1	Yellow	DC–	Aluminum, copper, magnesium alloys, nickel, mild steel and low alloy steels, stainless steel, and titanium
2% Thoriated	EWTh-2	Red	DC–	Aluminum, copper, magnesium alloys, nickel, plain carbon and low alloy steels, stainless steel, and titanium

Type of Electrode	*AWS Classification*	*Color*	*Polarity*	*Applications*
Zirconiated	EWZr	Brown	AC	Aluminum, copper and its alloys, magnesium alloys
Ceriated	EWCe-2	Orange	AC and DC–	Aluminum, copper, and magnesium alloys

No matter what kind of tungsten electrode you use, you're going to have to deal with the problem of *tungsten contamination,* which happens when the tungsten touches the filler rod or the molten weld pool. Believe me, it happens to almost all tig welders, especially as they're starting out.

A contaminated tungsten electrode can cause your welding arc to wander and can even cause the tungsten to split and shred into your molten weld pool. (Not a good thing, as you can imagine.) If your tungsten electrode touches your filler rod or weld pool while you're welding, you need to stop what you're doing and clean the tungsten by grinding or breaking off the contaminated area. Here's a quick look at those two options.

- **Grinding:** Grinding a tungsten electrode to remove contamination is the easiest option if you have a bench grinder with a diamond grinding wheel (see Figure 8-1). The bench grinder is stationary and easy to use, and the diamond grinding wheel is hard enough to make short work of the tungsten electrode. You can use a grinder with a stone or wheel that has a grit level of 120 or more.

  Grind your tungsten electrodes lengthwise so that the grind marks left on the electrode run the length of the electrode. If you don't grind in that manner, you can have problems with your arc when you go to weld again.

  Don't use the grinding wheel or stone that you use to grind your tungsten electrodes for grinding anything else because using it for other materials contaminates the tungsten. I always write "TUNGSTEN ONLY" on mine with a marker to remind me.

- **Breaking:** You can simply break off the contaminated tip of a tungsten electrode by gripping the electrode with a pair of sturdy pliers so that the contaminated tip (about ¼ inch usually does the trick) extends past the pliers. Then take a hammer and hit the tip squarely. It should break off and leave you with a relatively smooth electrode tip.

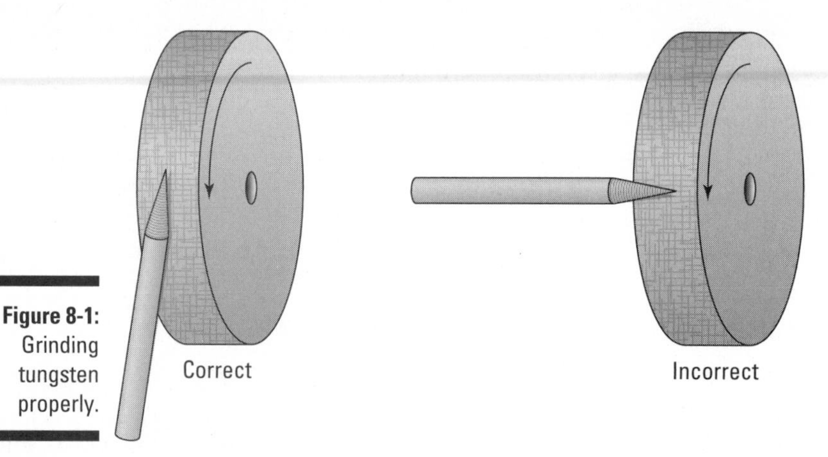

Figure 8-1: Grinding tungsten properly.

Don't even think about trying to take care of a contaminated tungsten electrode without wearing eye protection!

Making sure your shielding gas is set up correctly

Shielding gas protects your weld from the outside air and is a key component in any tig welding operation (see Chapter 7); you need to make sure you have your shielding gas ducks in a row before getting started with a tig weld. As I outline in Chapter 7, most welders use argon as their shielding gas for tig welding. You can get welding-grade argon at your local welding supply shop; it comes in a compressed gas cylinder.

The shielding gas is regulated during a tig weld by a *flow meter,* which ensures a constant gas flow and reduces the pressure of the shielding gas to a safe level. Your job is to set the flow meter so that it gives you the desired amount of gas flow, which is measured in cubic feet per hour (cfh). Too much or too little gas can be a problem. The former causes *porosity* (tiny holes) in your welds, and the latter can burn or deform your tungsten electrodes.

For most tig welding jobs, I recommend setting your flow meter at 12 to 15 cubic feet per hour as a starting point. That's usually an ideal range, but if you need to adjust the flow up or down, you can.

Make sure your shielding gas is hooked up correctly before you get started on a tig welding project. A hose should connect the shielding gas cylinder to your flow meter, and another hose should run from the flow meter to your welding machine. (In some cases, another hose connects from the welding machine through a *solenoid,* which is a switch that turns off and on to control gas flow. From the solenoid, the gas goes to the tig torch.) Your torch also has a gas valve, and you have to turn it on before you start welding and remember to turn it off when you're done.

There are a few variations to that kind of shielding gas setup. For example, sometimes a gas hose runs directly from the flow meter to the torch, but that's pretty uncommon.

Figuring out your tig filler rods

Some tig welds use *filler rods,* which provide extra metal to fill your weld. Chapter 7 covers the basics for tig welding filler rods, but here's one point that always bears repeating: Make sure the filler rod you're using is the same metal as the piece you're working on. Otherwise, you're creating alloys that you don't know the strength of.

The most common form of filler rod for tig welding is a 36-inch straight rod. These guys are pretty easy to use: You simply feed them into the molten pool with one hand while your other hand holds the tig torch. Filler rods come in boxes up to 50 pounds, and the type of rod is stamped or tagged on each individual rod, so you can easily identify the rods among the dozens of varieties.

Do what you can to keep your filler rods in pristine condition right up until the time you take them out of the packaging to use on a tig welding project. Rods that are contaminated by smoke, moisture, grease, or oil — even the oils from your skin — are more likely to produce faulty, weak welds. Rod condition is especially relevant with tig welding because it's such a precise, clean form of welding. If you have to handle filler rods, use gloves or make sure your hands are extremely clean. And if you need to tig weld with a filler rod that you think may have come in contact with grease or another contaminant, use a clean rag and a little alcohol or acetone to wipe down the rod before welding.

When tig welding with a filler rod, make sure you keep it inside the shielding gas zone. If the rod leaves the zone, it oxidizes; when you put the oxidized rod back in the molten pool, you add oxides to your weld, which weakens it.

If you find feeding a filler rod while you're tig welding difficult, consider buying flat rods next time. Many welders think the flat rods are easier to work with, and the cost is about the same as standard rods.

Note: Not all tig welds require filler rods. *Autogenous* welds use only the metal present in the welded piece. You can go the autogenous route on your tig welds, but you have to be really confident that the amount of metal present in the base metal you're working on will be enough to complete the weld without the molten pool burning through. (That's not a pretty sight.)

Matching Materials and Settings

Whenever I start talking about the versatility of tig welding, most people ask, "Just how many metals can you weld with tig welding?" My response is usually, "How much time do you have?" You really can tig weld a huge number of metals, including (but certainly not limited to) aluminum, bronze, copper, magnesium, nickel, gold, silver, steel, stainless steel, and titanium. And the list goes on and on.

But most tig welding beginners stick to the basics: steel, stainless steel, and aluminum. Those three metals are probably the most commonly used for new tig welders, and for that reason I present in this section a very straightforward breakdown of some of the settings (amperage, shielding gas flow, and electrode and filler rod diameter) that you should ensure when you're preparing to tig weld one of those three metals.

The settings I list in the tables in this section are general guidelines, so please don't regard them as the be-all and end-all. No two welding projects are ever exactly the same, so you may need to adjust some of these settings to make sure you're getting the best possible result.

Table 8-2 presents the settings you want to stick to for a few common steel tig welding scenarios.

Table 8-2 Common Settings for Tig Welding Steel

Thickness of Material (in in.)	*Amperage Required (in amps)*	*Electrode Diameter (in in.)*	*Shielding Gas Flow Rate (in cfh)*	*Filler Rod Diameter (in in.)*
1/16	90 to 115	1/16	15	1/16
3/32	105 to 120	1/16	15	1/16
1/8	115 to 135	1/16	15	3/32

Stainless steel is one of the metals that new tig welders love to tackle. If that's true for you, use the settings in Table 8-3 as guidelines to make sure you have your bases covered before starting a stainless steel tig welding job.

Table 8-3 Common Settings for Tig Welding Stainless Steel

Thickness of Material (in in.)	*Amperage Required (in amps)*	*Electrode Diameter (in in.)*	*Shielding Gas Flow Rate (in cfh)*	*Filler Rod Diameter (in in.)*
1⁄16	100 to 120	1⁄16	15	1⁄16
3⁄32	110 to 130	1⁄16	15	1⁄16
1⁄8	120 to 140	1⁄16	15	3⁄32

Use Table 8-4 as an at-a-glance reference tool for tig welding aluminum.

Table 8-4 Common Settings for Tig Welding Aluminum

Thickness of Material (in in.)	*Amperage Required (in amps)*	*Electrode Diameter (in in.)*	*Shielding Gas Flow Rate (in cfh)*	*Filler Rod Diameter (in in.)*
1⁄16	70 to 90	1⁄16	15	1⁄16
1⁄8	140 to 160	3⁄32	17	1⁄8
3⁄16	190 to 220	1⁄8	21	5⁄32

Getting a Handle on Using Your Tig Torch

As with the other arc welding techniques, one key to a good tig weld is the proper handling of your torch. The following sections show you how to strike an arc on your tig torch and maintain the right grip.

Don't attempt any part of the tig welding process unless you're very familiar with welding safety. If you need a refresher, check out Chapter 3.

Choosing an electrical current and striking the arc

Striking an arc is a critical first step in the tig welding process. It's a skill that you want to work on so that you can do it quickly and effectively as you continue to improve as a welder, and I walk you through it in this section.

But before you strike an arc for tig welding, you need to decide which electrical current you want to use. Here are your choices:

- **Direct current electrode negative (DCEN or DC–):** Also known as straight polarity, this option produces the greatest amount of heat in the metal you're working on. Because of that, DC– is the most popular choice for tig welding. The major drawback to using DC– is that it doesn't offer any cleaning action on the surface of the work.
- **Direct current electrode positive (DCEP or DC+):** In this mode (also referred to as reverse polarity), most of the heat is on the tungsten electrode, so you tig weld with it when you're working with really thin pieces of metal only.
- **Alternating current (AC):** When you use AC while tig welding, half the heat produced is located in the piece you're welding, and half is in the tungsten electrode. AC is a great option when you're tig welding aluminum or magnesium because it cleans the work during the welding process and maintains a stable arc.

If you're tig welding aluminum with AC, you must run your machine on continuous high frequency, which I cover later in this section.

After you decide which electrical mode you want to use for your tig welding project, you're ready to strike an arc. You can do that by using one of two methods: scratch starting or remote control.

Before striking the arc with either method, you must have a lens in your welding helmet that protects your eyes but still allows you to see the weld area. For tig welding with less than 50 amps, I recommend a #8 shade. For 50 to 150 amps, go for a #11 or #12 shade. If you're trying to decide between two different shade numbers, always go with the darker of the two. You can always go to a lighter shade if you know your eyes will be protected and want a better view of the weld area.

- **Scratch starting:** *Scratch starting* involves starting the machine and then touching the tungsten electrode against the piece of metal you're welding in order to strike the arc. Many tig welding machines have made this method somewhat obsolete (more on that in the following bullet). I definitely don't recommend using the scratch start method when you're tig welding stainless steels or alloy steels.
- **Using a remote control device:** A remote control device allows you to strike an arc without touching the electrode to the piece of metal you want to weld. You simply place your electrode about ⅛ inch above the spot where you want to start tig welding, and then turn on the current.

  Several types of remote control devices are available. One of the most prominent is a foot pedal connected to your tig welding machine. When you press the pedal down, it turns on the welding current and usually also starts the shielding gas flow (as well as the cooling water flow through your tig torch, if your torch is water cooled). Some foot pedals also have a control that allows you to gradually decrease the amount of welding current produced by your machine. That makes it easier to fill the end of the weld, which lessens the chance of the *crater cracks* that form when you don't pause at the end of the weld and fill the crater. That's especially useful when you're tig welding aluminum because it shrinks faster than other metals and is therefore more prone to crater cracks.

Another type of remote control device is a simple button. You hit the button to start the welding current, gas flow, and water flow, and you hit the button again to turn everything off.

Remote control devices that allow you to strike a tig welding arc use a high-frequency current generated by your tig welding machine. Most tig machines that come with this option have a switch that allows you to choose between *high frequency start only* and *continuous high frequency.* If you set the switch on high frequency start only, the high frequency current shuts itself off after you strike the arc. Continuous high frequency current continues running. You can read more about high frequency modes in Chapter 7.

If you're using DC, the electrode has to touch the work to start the arc. Make contact between the electrode and the work and then turn on the current. That strikes the arc, and you can then move the electrode up about ⅛ inch off the parent metal to create the weld pool.

To stop the arc regardless of what current you're using, lift the electrode to the horizontal position, or at least two inches away from the work.

Get a grip: Holding your tig torch correctly

Holding your tig torch the right way when you're tig welding is important — an improper grip can cause not only poor welds but also unnecessary stress on your hand and arm.

To get the most precise grip on a tig torch, grab the torch as you would a pen halfway up the part of the torch that the hoses connect to.

When you're using the tig torch, keep it at a 5- to 15-degree angle. That allows you to see the weld area clearly but still get adequate shielding gas flow to the weld area. If you increase the angle, the torch won't be able to cover the area with shielding gas, and a tighter angle will likely make it difficult to see what you're welding (never a good thing).

Giving Tig Welding a Try

After you have all the preparations taken care of and all your safety precautions straight, you can try out a tig weld. Hopefully, this experiment will be the beginning of a long, productive relationship between you and tig welding, so don't get too worried if your first tig weld doesn't win any prizes or set any records.

In the following sections, I walk you through the steps for making a tig weld in the flat position. Then I tell you how you can continue to try tig welding on a few different types of joints.

Tackling the first weld

Here's what you need (in addition to the tig equipment mentioned in this chapter and Chapter 7) for the first weld:

- A 2-inch-x-5-inch piece of 16 gauge mild steel 1⁄16 inch thick. Make sure the piece is clean and free of scaling. (You may need to grind it a little.)
- An RG60 filler rod with a diameter of 1⁄16 inch.
- A 2-percent thoriated tungsten electrode (see Table 8-1 earlier in the chapter).

Here are a few final checks you should make before you get started:

- Make sure all of the electrical connections are tight.
- Make sure you have the proper size of tungsten electrode, and check to be sure that it's tight by pressing the tungsten against your welding table. If it moves, tighten the *collet holder* (which holds the tungsten tight during the welding process) a little. (Don't overtighten, or you'll strip the threads.)
- Make sure you have the right shielding gas cup size (see Chapter 7).
- Make sure your piece of steel is lying flat on your welding table so that it won't move around or fall off while you're working on it.

All set? Great! Time to get started. For this practice, you just work on welding a line down the middle of the 5-inch length of your piece. (You're not joining any pieces on this first go-round.) Here's how you do it.

1. **Set the welding machine for the amperage required to work on a piece of metal with a thickness of 1⁄16 inch — about 75 amps.**
2. **Select the high frequency start only option on your welding machine.**
3. **Select DC– (straight polarity).**

4. **Turn on your shielding gas and adjust the flow to 15 cubic feet per hour.**
5. **If you have a water-cooled tig torch, turn on the water flow.**
6. **Strike the arc by holding the electrode ⅛ inch above the piece to be welded, turning on the welding current, and pushing down on the foot pedal.**

   Your machine may use a different method for striking the arc; read more about that in "Choosing an electrical current and striking the arc" earlier in this chapter, and if in doubt, check your machine's instructions.
7. **Hold the tig torch so it points in the direction you want to weld.**
8. **Move the torch slowly toward the metal piece and increase the pressure on your foot pedal until the pool has a diameter of about 2½ times the diameter of your filler rod.**
9. **Using your free hand, grab your filler rod so that 12 to 15 inches of the rod extend past your forefinger.**
10. **Dip the filler rod into the molten puddle and then pull it out of the puddle slightly; move the torch in the direction of the weld.**
11. **Add more filler rod and repeat until you reach the end of the piece of steel.**

If you contaminate the tungsten electrode by touching it to the mild steel piece or the filler rod, you must stop and clean the tungsten by using the steps I describe earlier in this chapter.

Trying a butt joint

A *butt joint* is a weld between two pieces of metal lying in the same plane. If you take a couple of pieces of metal, lay them out flat on your welding table, and then slide them together, a butt joint forms where the two pieces touch. You can see some examples of butt joints in Figure 8-2: Figure 8-2a shows a butt joint in aluminum, Figure 8-2b features a stainless steel butt joint, and Figure 8-2c illustrates a mild steel butt joint.

For the butt joint practice in this section, you need two pieces of 2-inch-x-5-inch, 1⁄16-inch-thick mild steel. You also need the same kind of filler rod and tungsten electrode I describe in the preceding section. Lay the pieces of steel flat on your welding table so that the two 5-inch edges are touching and then take the following steps:

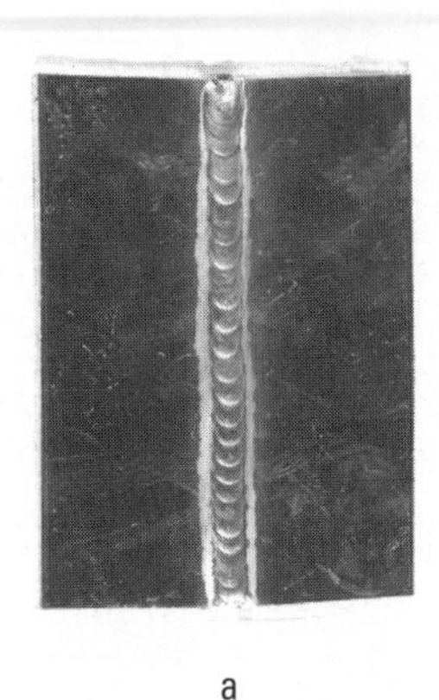

a b c

Figure 8-2: Aluminum (a), stainless steel (b), and mild steel (c) tig welded butt joints.

1. **Hold your tig torch at between 65 and 75 degrees to the surface of the steel.**
2. **Strike an arc (see "Choosing an electrical current and striking the arc" earlier in the chapter).**
3. **Begin forming the weld puddle with the arc; when the puddle reaches a diameter 2½ times the diameter of your filler rod, move the arc away and start adding your filler rod to the weld puddle, keeping the rod at an angle of 20 to 25 degrees at the front edge of the puddle.**
4. **As the filler rod melts into the puddle, keep adding more filler rod to the leading edge of the weld.**

   If the puddle begins to get less fluid, move the filler rod away and use the electrode to heat the puddle back up so it's nice and molten.
5. **Repeat Steps 3 and 4, traveling along the length of the joint, until the weld is complete.**

Welding a lap joint

A *lap joint* is a weld made joining two pieces of overlapping metal. It's one of the strongest joints, and one that you'll probably want to master as you continue to expand your tig welding horizons. Figure 8-3a gives you an example of an aluminum lap joint, and Figure 8-3b shows one in stainless steel. In Figure 8-3c, you see a mild steel lap joint.

To make a strong lap joint, you have to make sure that the two pieces of metal overlap by at least 2½ times the thickness of the thickest piece.

Figure 8-3: An aluminum (a), stainless steel (b), and mild steel (c) tig welded lap joint.

To practice tig welding a lap joint, you need two pieces of 2-inch-x-5-inch mild steel 1⁄16 inch thick, as well as the same electrode and filler rod from the exercises in the preceding sections. Lay the pieces on your welding table so they overlap by ½ inch. (I know that's more than 2½ times the thickness of the pieces, but for this practice exercise it makes things easier.)

1. **Strike an arc and begin by forming a puddle on the bottom plate, close to where the top plate is overlapping; after the puddle forms, shorten your arc to no more than ⅛ inch.**
2. **When the puddle forms a *V* between the two plates, add your filler rod to the center of that *V*, dipping the rod slowly in and out of the puddle every 3⁄16 of an inch as you work your way down the joint at a steady speed of travel.**

   If you keep up a steady speed of travel and dip in your filler rod as noted here, you should end up with a uniform bead and a strong weld.

Making a T joint

A *T joint* is exactly what it sounds like: a weld that joins two pieces of metal that form a *T* shape at a 90-degree angle. You can see aluminum, stainless steel, and mild steel *T* joints in Figures 8-4a through c, respectively.

Figure 8-4: An aluminum (a), stainless steel (b), and mild steel (c) *T* joint (C)

To tig weld a *T* joint, use two pieces of 2-inch-x-5-inch, 1⁄16-inch-thick mild steel. You also use the same electrode and filler rods as for the practice welds in the preceding sections. Lay one piece flat on your welding table and use a pair of pliers to hold the other piece at a 90-degree angle at the end of the piece on the table. Then follow these steps.

1. **Start by striking an arc and tack welding the two plates together with an autogenous weld.**

   No need to use a filler rod for a quick tack; just fuse them together autogenously to keep them in place so that you can go back and make your finished weld later.

2. **Still working autogenously, use your torch to start a weld puddle at the end of the joint closest to you on the plate that's laying flat on the table.**

3. **Turn your wrist to rotate the torch until it's at a 45-degree angle between the two pieces of material to be welded.**

4. **Move the torch along the joint, and begin adding in your filler rod every 1⁄4 inch down the weld until you get a uniform bead with the weld deposited equally between the two joints.**

5. **When you reach the end of the joint, be sure to fill the crater.**

   You're done!

Part III
Discovering Mig Welding

The 5th Wave	By Rich Tennant

"Should I know anything about arc welding?"

In this part . . .

How can I put this delicately? Mig welding is, well, easy. I know, I know, that's how rumors get started. But mig really is easy, and you should take advantage. (My apologies if anyone reading this happens to be named Mig.)

The mig welding process has really gained popularity in recent years because it's the kind of technique that you can successfully use without putting in hours and hours of practice. A lot of the process is automated, and as a result it's not only easier but also faster than the other types of welding. It's not perfect — I tell you about the drawbacks as well as the advantages of mig welding in this part — but I think it's something you're going to want to know how to do. To that end, I also help you practice a bit of mig welding in this part as well.

Chapter 9

Understanding the ABCs of Mig Welding

In This Chapter

- Exploring mig welding's basics, advantages, and disadvantages
- Understanding mig equipment and supplies
- Figuring out important settings for a few mig welding scenarios

Mig welding is an arc welding process (like stick or tig welding — see Chapters 5 through 8) that first started showing up in the late 1920s, and its popularity quickly skyrocketed. Why? Mig welding is fast and easy.

I know what you're thinking: "If mig welding is fast and easy, why the heck do you spend hundreds of pages talking about any other kind of welding in this book?" Compared to other types of welding, mig can be quite a bit faster. It's relatively easy, but it's certainly a lot easier if you've had some experience with (or you're familiar with) other types of welding. And despite the fact that it's easier to pick up and quicker than some other welding processes, it's still not the best choice in every welding situation.

Either way, if you're interested in getting into welding, you're going to want to know how to mig weld, and I start you on that path in this chapter. Read on to begin wrapping your brain around all of the basics of mig, including how it works and what supplies are involved.

Note: *Flux core arc welding* (also known as FCAW) is very similar to mig welding, but it doesn't involve shielding gas. Instead of gas, a section of flux in the middle of the electrode wire shields the molten puddle after welding, and it has to be removed upon completion of a weld. Mig welding is much more common — especially among beginning welders — so I don't cover it in detail in this chapter.

Understanding How Mig Welding Works

Two of the most important characteristics of mig welding are the use of a continuously fed supply of electrode wire (often called simply *wire*) and the need for a *shielding gas* to protect the weld pool from the atmosphere — atmospheric gases can make your welds brittle, porous, or just plain weak. You feed the electrode into the mig welding machine, which directs it down into the mig welding gun (the part you hold with your hand). In the *gun,* an arc melts the electrode and allows you to deposit it on your *parent metal* (the metal you're making the weld on) to make a weld. The gun also directs a shielding gas flow (usually argon) over the weld pool.

Like the other major welding processes, mig is a welding technique that goes by many names. You may see mig referred to as *gas metal arc welding* (GMAW), *flux cored arc welding, spray arc welding,* or *short circuit welding,* to name just a few.

Mig welding machines can use electricity in a few different ways to transfer the metal from the electrode into molten metal that you can use to weld.

- **Short circuit transfer:** This option is the most common transfer process. It uses low amperage, which allows the liquid metal at the tip of the electrode to transfer to the molten pool by direct contact. As a result, it produces a very low amount of *spatter* (small pieces of metal that stick to the parent metal surface). Short circuit transfer works well in all positions on low carbon steel, alloy steel, and stainless steel.

  Mig welding machines that use short circuit transfer make a loud buzzing sound during operation, so don't think you're being attacked by a swarm of killer bees when you start using a short circuit transfer machine. That sound is normal.

- **Spray arc transfer:** As the name implies, *spray arc transfer* happens when the metal at the end of the wire electrode gets melted into small droplets and then is sprayed from the wire to the base metal. The arc propels the droplets at a high speed in the direction that the electrode wire is pointing. This method is used mostly in the flat position, and it has three major requirements:
  - You have to use it with direct current electrode positive (DC+) polarity. (Check out Chapter 5 for more on polarity.)
  - You have to keep your voltage above a certain level. (You can read more about those requirements in the user manual for your spray arc transfer machine.)
  - You have you use a shielding gas that's at least 80 percent argon.

- **Pulse transfer:** This kind of transfer is somewhat similar to spray arc transfer, but it uses a constant pulsing of current that makes the metal move from the electrode wire to the weld puddle in a more consistent manner than the spray arc method does. The pulsing allows you to accomplish the same effect as spray arc transfer but with lower amounts of electrical current and therefore lower heat. You can weld in several different positions with pulse transfer, and you can also weld some very thin materials, such as aluminum and all steels down to 20 gauge in thickness.
- **Globular transfer:** In this process, a molten ball forms on the end of the electrode wire until it's two or three times the size of the wire. The ball then falls off and becomes part of the weld pool. Shielding gas is really important with this type of transfer because the molten metal falls through the shielding gas and attaches itself to the parent metal. Globular transfer can also create more spatter than the other types. Not many mig welding machines use globular transfer, but it can be useful when you're working with thin materials on a really low amperage setting.

Mig welding can involve several variations that depend on the shielding gas, the type of metal to be welded, and the type of metal transfer that takes place during the weld. I touch on a few of those variations in this chapter, but for the most part I stick to the basics.

Do the robot (weld): Robotic mig welding

Robotic welding completely automates the welding process. Mig welding robots work best when a large number of pieces need to be welded quickly and in exactly the same way every time. As you would probably imagine, the robots are used most often in controlled environments. (And even then, a qualified welder often needs to also be on site, checking to make sure that the robots do the work correctly and making adjustments and fixes when necessary.)

Some mig welding applications in industry have been successfully robotized. For example, the auto industry has used robotic mig welding machines for more than 25 years. Roughly 60,000 welding robots are in use in North America today. The opportunities exist for more, but the cost of the machines is prohibitive for many businesses and applications.

Considering Mig Welding's Advantages and Limitations

As I say in the introduction to this chapter, mig has risen in popularity in recent decades because it's relatively fast and easy. But you should know about some other advantages and keep some limitations in mind as well. Here are the advantages:

- **It's fast.** Okay, so this one isn't "another" advantage, but it bears repeating. If you had a race between mig and all the other major welding processes under identical conditions, mig welding would win handily, primarily because of its continuously fed electrode. Can't beat "continuous" when it comes to speed.
- **It's easy.** When I say "easy," I don't mean that you can take 15 minutes this weekend and teach your 8-year-old daughter how to mig weld. (You shouldn't do that anyway, of course, because welding and kids go together about as well as weights and skates.) But because several of the processes that take place during a mig weld are automated — again, the continuously fed electrode is at the top of this list — it's easier to pick up and succeed with quickly than, say, tig welding.
- **It eliminates almost all need for cleanup time.** Mig isn't quite as tidy as tig welding, but it's still pretty clean when you compare it to stick. (See Chapters 5 and 7 for stick and tig cleanliness, respectively.) Unlike the mess generated by stick welding, mig doesn't involve flux or create much *slag* (debris).
- **It allows you to make larger passes with a single weld.** These longer passes are another reason that mig is faster than the other welding processes. With mig welding, you're able to deposit more metal with each pass, and that means you can get away with fewer passes when you're working on a big joint.

  The width of the mig weld pool is determined by the *speed of travel* (how fast you're welding).
- **It cuts down on the number of starts and stops.** When you use a continuously fed electrode, it definitely helps to limit the number of times that you have to stop welding, adjust or get a new electrode, and then start welding again. It may not seem like a lot, but when you think about how much time you can lose stopping and starting over the course of a project (or a career!), it can really add up.
- **It eliminates electrode stub loss.** One problem that really bothers me about stick welding is that, no matter how well I weld, I still can't use the last few inches of an electrode. (It's unsafe, ineffective, and pretty much impossible to use a whole electrode when you're stick welding.) It just seems so wasteful! Well, with mig welding, that wastefulness isn't an issue because you *can* use the whole electrode.

- **It works on metals in a wide range of thicknesses.** Mig allows you to achieve good *penetration* (weld depth) and still produce a smooth, strong weld bead.

The amount of current you use determines the penetration you can achieve with mig welding.

Mig welding isn't all upside. Here are some of the technique's key limitations.

- **Mig welding equipment is costly.** That's especially true when you compare the cost of mig gear to the cost of stick welding equipment. Mig equipment — particularly the machine — is quite a bit more complex than what you use for stick welding, and it costs more to manufacture. Who pays for that increased manufacturing cost? You do, of course!
- **Mig equipment requires more maintenance than other types of welding equipment.** That usually means more down time (while you're adjusting your equipment or waiting for the guys at the welding shop to fix it) and more cost.
- **Mig welding outside is difficult.** Because you have to use a shielding gas, any kind of a breeze can disrupt your ability to mig weld effectively. You can accomplish mig weld outside, but it has to be a very still day (no wind), or you have to use a really good windscreen if you're going to be successful. (Chapter 4 gives you more details on windscreens and other supplies.) Another hindrance: Mig welding equipment isn't very portable, so getting it in and out of your shop on a regular basis may be problematic. If you're looking for maximum portability, you may want to look to another type of welding.
- **Mig guns are kind of big.** A mig welding gun has a lot going on inside, and as a result it's a bulkier piece of equipment than what you may expect to use for, say, stick welding. This limitation isn't the kind of disadvantage that's going to make you want to throw down your mig gear and swear off mig welding for the rest of your life, but it can be a little bit of a pain sometimes. For example, sometimes manipulating a mig welding gun in a tight space can be difficult.

Bringing out the Big Guns (And Other Mig Welding Equipment)

Mig welding requires some specialized equipment, and in this section I walk you through the details of what you need to get started with mig. Note that I cover only mig-specific *equipment* in this section; if you want to know more about the critical supplies you need for mig welding (electrode wire and shielding gas), see the relevant sections later in the chapter. To get a peek at what a bona fide mig welding setup looks like, check out Figure 9-1.

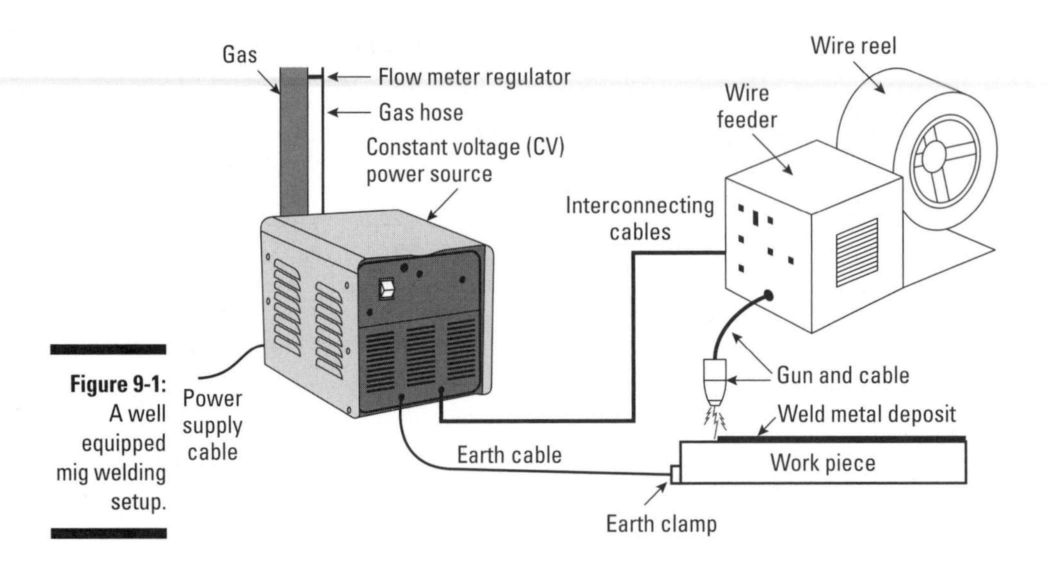

Figure 9-1: A well equipped mig welding setup.

Mig welding machines

Mig welding machines come in several different varieties, and the differences from one machine to another are based mostly on the amount of voltage that it can put out. You can plug some smaller mig machines into a standard 110-volt electrical outlet (like the smaller ones in your house, such as one you'd use for a toaster), and others require a bigger 220-volt power supply. (A 220-volt outlet is the kind you need to plug in an electric clothes dryer.)

Most mig welding is done with direct current (DC), or DC+, to be more specific. Make sure any mig machine you're looking at buying produces direct current. (To be honest, these days you'd probably be hard pressed to find a mig machine that didn't have a DC option, but double-checking never hurts.)

Mig welding machines vary in price. You can get a small, less powerful machine for $300; if you want to be able to mig weld heavy-duty metal (½-inch steel, for example), you're looking at a machine closer to $1,000.

Mig welding guns

Happiness is a warm mig welding gun. No, really! A mig welding gun is part of what makes mig welding such a user-friendly welding process. It delivers the electrode wire, welding current, and shielding gas all to the same place at the same time.

Your mig welding gun is attached to a power cable leading to your mig welding machine. It's also attached to a shielding gas hose leading to your shielding gas flow meter. Mig guns have a trigger that start and stop the welding process easily and quickly. You just pull the trigger and all at once you have shielding gas flow and a welding arc met with a steady flow of electrode wire. It doesn't get much easier than that!

Electrode wire feeders

Mig welding requires an *electrode wire feeder,* which adds a steady supply of electrode wire to the weld.

You have to match the electrode wire feeder to the machine. For example, you can't put a 110-volt wire feeder on a 220-volt mig machine. You also can't put a wire feeder made by one welding equipment manufacturer on a machine made by a different manufacturer.

You have a few options when it comes to choosing an electrode wire feeder. Here's the scoop on all of them.

- **Push feeders** have rollers that are securely clamped against the wire to push it through to the gun. A groove in the roller keeps the wire from slipping. Some rollers are *V*-shaped, some are *U*-shaped, and some are notched, depending on the wire that's being pushed. You can change rollers when you need to change your wire. Push feeders have an adjustment so you can change the amount of pressure on the wire and keep it from slipping.
- **Pull feeders** use a small motor located in the welding gun to pull the wire. You use these systems on relatively soft wire (such as aluminum). A major disadvantage to a pull feeder is the weight they add to the gun. (It can get heavy if you're working on a long project.)
- **Push-pull feeders** are a combination of a push feeder and a pull feeder: A wire feeding unit on top of the welding machine pushes the wire, and the gun holds a small motor with a set of rollers for pulling. The unit pushes and the rollers pull all at the same speed. Push-pull feeders are great for moving any type of wire over long distances. However, because they're more complex, these types of feeders are more expensive than your other options.
- **Spool guns** are self-contained drive systems consisting of a wire supply and a set of rollers, both of which are located in the gun. They allow you to move around freely while dragging only the power lead and shielding gas hose. Because the wire has to travel only a short distance, spool guns are really useful for soft wires like aluminum.

Sifting through Shielding Gases for Mig Welding

As I note earlier in the chapter, shielding gas is critical for mig welding because it keeps the weld puddle from becoming contaminated by the atmosphere. You want the shielding gas to flow over the work area at a nice, steady rate.

The setup for shielding gas in a mig welding operation is fairly straightforward. The shielding gas is in a pressurized cylinder, and you regulate it down to a useable working pressure with a flow meter. The gas is then fed out to the mig welding gun and through a nozzle onto the molten pool. Pretty simple, right?

The not-so-simple part is choosing a shielding gas for your mig welding projects. You have a number of shielding gas options to choose from, and lots of criteria to consider when making your choice, including depth of penetration, porosity prevention, welding speed, and of course cost. Here's some more information to help you figure out what will work best for your mig jobs.

- **Argon and argon mixes:** You can use pure argon as a shielding gas, or you can use argon that has been mixed with another gas. One important positive for argon is that it's heavier than air (unlike helium — see the following bullet), so you can use less of it while you're welding because it flows from the nozzle directly down onto the weld area. Argon is also great if you want a nice, quiet arc and a low amount of spatter. Expect to pay about $100 dollars for a large argon or argon mix cylinder.
- **Helium and helium mixes:** Pure helium and helium gas mixtures can also serve as shielding gas. The advantage to helium is that it can help you produce wider welds, so it's useful when you're welding heavy sections of metal. The major downside is that helium is lighter than air, so it has a tendency to float away from the weld puddle. Because of that characteristic, you have to increase the flow rate and use quite a bit more gas than you would with, say, argon to get the job done. Helium is more expensive than argon; it costs about $10 more for a cylinder (244 cubic feet).
- **Carbon dioxide (CO_2):** Carbon dioxide offers a nice, wide weld pattern and deep penetration. The downside is that it often results in the production of quite a lot of messy weld spatter. It's the cheapest of the three shielding gas options (about $12 for a 244-cubic-foot cylinder.)

So how do you know what kind of gas to get for your new project? Here are my recommendations for the types of shielding gas you should use for mig welding various kinds of metal in different types of transfer modes. (Flip to "Understanding How Mig Welding Works" earlier in the chapter for more on transfer modes.) Note that some of my suggestions call for gas mixtures, which the folks at your local welding supply shop can get for you.

- **Most aluminum alloys:** 100 percent argon
- **Alloys of aluminum and magnesium:** 75 percent helium/25 percent argon
- **Short circuit transfer on low carbon steel:** 75 percent argon/25 percent carbon dioxide
- **Short circuit transfer on mild steel:** 90 percent argon/10 percent carbon dioxide
- **Short circuit transfer on deep penetration welds used for pipe welding:** 50 percent argon/50 percent carbon dioxide
- **Short circuit transfer on deep penetration welds used for welding heavy wall pipe:** 25 percent argon/75 percent carbon dioxide
- **All transfer modes on carbon steel, low alloy steel, and stainless steel:** Argon with 5 to 10 percent carbon dioxide and 1 to 3 percent oxygen
- **Short circuit transfer on high strength steel in all positions:** Helium with 20 to 35 percent argon and 4 to 5 percent carbon dioxide

Taking a Look at Electrode Wire

One of the things that make mig welding unique is the use of electrode wire, which automatically feeds into the weld area. Mig welding electrode wire is tubular and solid, although it does have various powdered materials incorporated in its core. It's also smaller than what you'd use for, say, stick welding. The key measurement is the diameter; diameters of mig electrode wire range from .023 inches to .045 inches. Mig wire is appropriate for both single-pass and multiple-pass applications.

The size you choose for each mig welding project is based on the thickness of the metal you're welding and also the position in which you'll be welding. If you're going to be doing out-of-position mig welding, you need to get smaller electrode wire. Because electrode wire is fairly thin, and because mig welding requires high levels of electric current, electrode wire melts off at a rate of anywhere between 40 and 900 inches per minute!

When you're buying mig electrode wire, always remember that the metal must match the metal you're going to be welding. Don't buy stainless steel wire if you're going to be working on an aluminum mig welding project.

Most welding supply stores and other types of retailers sell mig wire by weight, either on spools or in drums. You can buy mig electrode wire in a drum that contains anywhere from 300 to 1,000 pounds of coiled wire! Common weight increments are 1 pound, 3 pounds, 10 pounds, 25 pounds, 44 pounds, and 50 pounds. As you can imagine, you can get more wire for your money if you go for a larger spool or drum, but the catch is that you have to be very confident that you want to buy a large amount of the same type of wire. If you're going to be doing a small steel project followed by a small aluminum project and then maybe a small stainless steel project, you probably don't want to buy 500 pounds of aluminum wire.

One of the most frequently used types of mig wire is carbon steel wire, so Table 9-1 gives you the details on some common types of carbon steel wire.

Table 9-1 Common Carbon Steel Wire Classifications

Wire Classification	*Uses*
ER70S-2	Welds all grades of carbon steel and carbon steel that's rusty
ER70S-3	Used for welding appliances and farm equipment (good general repair wire)
ER70S-4	Used for welding boilers and pressure vessels, and for other structural steels
ER70S-5	Welds dirty and rusty steel best
ER70S-6	Welds most carbon steels and works well with carbon dioxide shielding gas
ER70S-1B	Welds low alloy steels that can be slightly rusty
ER70S-7	Creates high-quality, sharp-looking welds (minimizes porosity and weld defects) on steel

Adjusting Mig Equipment to Suit Your Mig Welding Project

A lot of variables come into play when you're mig welding. You have to figure out your electrode wire size, wire feed speed, voltage, and shielding gas type, and all those factors have to jibe with the thickness of the metal you're welding.

Table 9-2 helps you to figure out some of those details by listing the wire feed speed (first number) and voltage (second number) recommended for each parent metal thickness and wire size combo. Wire feed speed numbers are in inches per minute, and voltage numbers are in volts. This table applies to steel parent metals only.

For example, if you're mig welding on a piece of metal that's 3⁄16 inch thick and you're using a mig electrode wire size of .030 inch, you would want a wire feed speed of 280 inches per minute and voltage of 19 volts if you were using a shielding gas mixture of 75 percent argon 25 percent carbon dioxide. If your shielding gas for that project was 100 percent carbon dioxide, you'd want to keep your wire feed speed at 280 inches per minute but increase your voltage to 20 volts.

Table 9-2 Feed Speed and Voltage Settings for Common Mig Welds

		Metal Thickness								
Wire Size	***Gas Composition***	***22 Gauge***	***20 Gauge***	***18 Gauge***	***16 Gauge***	***14 Gauge***	***12 Gauge***	***10 Gauge***	***3⁄16 in.***	***1⁄4 in.***
.023 in.	75% argon, 25% CO_2	70/18	120/15	140/16	190/16	260/17	330/18			
	100% CO_2	n/a	120/18	140/19	190/19	260/20	330/20			
.030 in.	75% argon, 25% CO_2		80/15	90/16	130/16	180/17	230/18	280/19	330/19	
	100% CO_2		80/18	90/19	130/19	180/20	230/21	280/22	330/21	
.035 in.	75% argon, 25% CO_2			90/16	120/16	160/16	200/17	240/18	280/18	320/19
	100% CO_2			90/19	120/19	160/19	200/20	240/21	280/20	320/22
.045 in.	75% argon, 25% CO_2			80/16	100/16	120/17	140/18	160/19	180/19	200/20
	100% CO_2			80/19	100/19	120/20	140/21	160/22	180/22	200/23

Chapter 10

Practicing Mig Welding

In This Chapter

- Getting ready to mig weld
- Giving mig welding a try with some practice exercises
- Looking out for common mig weld defects

If you've heard of mig welding (or you've read Chapter 9), you know that it's a fast and relatively easy arc welding method. The continuously fed electrode wire sets mig apart from the other arc welding techniques (stick and tig). You can pick it up and understand it pretty quickly, and the quality of the welds that you can achieve with mig welding is extremely high.

New welders always want to try mig welding after they hear about it, and I'm not one to disappoint — this chapter is all about how you can get started actually practicing mig welding. I walk you through the important steps required to prepare for and make a mig weld. I round out the chapter with an explanation of some common problems that welders (especially new welders) see in their mig welds so that you can get some idea of what may be going wrong if your work isn't as good as you want.

Preparing to Mig Weld

Although mig welding is fast and easy, it's not a magical welding process that allows you to just walk into the shop, pick up some equipment, and make perfect welds within 20 seconds. Because mig equipment is a little more sophisticated and complex than some of the other pieces of equipment in your welding shop, the preparation for mig welding generally takes a little longer than what you may expect for the other arc welding processes. I'm not talking about hours and hours of tedious setup, so don't let it scare you off; just know that proper preparation is the key to good mig welds. The following sections clue you in on how to get ready to make a mig weld.

Getting the equipment ready

As with any kind of welding, you have to make sure your equipment is in order before you can get started. The following steps give you the lowdown on setting up your mig machine. (If you're not familiar with mig equipment, you may want to look at Chapter 9 — if you haven't already — to get a handle on the items I talk about in this section.)

Before you operate mig equipment, make sure you've read and understand all the relevant instruction manuals. Also check to see that the machine is assembled according to the manufacturer's specifications. Finally, check all the connections, hoses, and so on to ensure that they're in good shape.

Don't operate any equipment without the proper safety gear. Check out Chapter 3 for the rundown on safety equipment and precautions.

1. **Install a shielding gas supply cylinder on the back of your mig welding machine, and make sure it's chained securely before removing the cap.**
2. **Stand to one side of the cylinder and open and close the valve quickly to blow out any dirt that may be trapped in it.**
3. **Install the flow meter on the cylinder.**
4. **Attach the hose to the regulator, and attach the gas end connecter to the electrode feed unit.**
5. **Install the welding wire on the holder, making sure you line up the pin on the back of the spool and double-checking that the wire matches the size required for the job you're planning to do.**
6. **Make sure the *liner,* which carries the wire from the rollers to the contact tip, matches the wire size you are welding with.**
7. **Make sure the power is off and then attach the welding cables.**
8. **Attach the electrode lead to the positive terminal and the ground lead to the negative.**
9. **Attach the welding gun (complete with gas diffuser, contact tip, and shielding gas nozzle) into the electrode feeding unit or the wire feeding unit.**
10. **Cut off the end of the electrode wire that's wrapped around the spool of the electrode feed wire, holding onto it so that it doesn't unwind; separate the rollers and push the wire through the guide and then between the rollers and into the liner.**

    You'll probably have to read the instruction manual for your electrode wire feeder in order to do this step correctly.
11. **Apply some tension to the rollers and set the wire feed speed at 33.**

    That setting is a good starting point — not too fast, not too slow.

12. **Press the gun switch or the jog button so that the wire starts feeding smoothly until it comes out of the contact tube.**

    When you see it coming out of the contact tube, you're definitely on the right track. If it doesn't, you don't have enough pressure on the *drive rolls* (two round steel wheels with a groove in the center that are attached to the wire feeder and turn at different speeds to push the wire through the liner to the welding puddle).

    When dealing with your welding gun, you also want to be aware of *stickout,* the distance that the electrode wire extends past the gun's contact tip. It controls penetration and fusion. You usually use stickout of ⅜ to ¾ inch.

13. **Open the valve on the shielding gas cylinder all the way and turn on the machine.**
14. **Slowly turn in the adjusting screw and watch the ball rise in the tube as the gas begins flowing.**

    The tube has different scales for different gases; make sure you're reading the one that matches the gas you're using.

15. **Release the pressure on the drive rolls and pull the trigger switch.**
16. **Set your shielding gas flow rate.**

    The rate depends on factors like the type of material you're welding and the thickness of that material. To give you an idea of how flow rates can change, have a look at Table 10-1, which gives info for varying thicknesses of mild steel and different combinations of shielding gases. (Head to Chapter 2 for more on mild and other kinds of steel.)

Table 10-1 Some Shielding Gas Flow Rates for Mild Steel

Thickness	*Flow Rate for 100% CO_2 (in cfh)*	*75% Argon/25% CO_2 (in cfh)*	*98% Argon/2% Oxygen (in cfh)*
16 gauge	20	25	25
14 gauge	20	25	25
⅛ in.	22	25	25
3⁄16 in.	22	30	30
¼ in.	24	35	35

17. **Reset the pressure on the drive rolls.**

    You have to reset the drive rolls' pressure because you released it to set the shielding gas pressure in Step 15.

18. **Attach the ground clamp to your welding table or the piece you're going to be working on.**

Setting the wire feed speed and voltage

The continuously fed electrode wire (one of the hallmarks of mig welding) means that a lot of the welding process is automated, but you have to have a few key settings on your mig machine (electrode wire feed speed, amperage, and voltage) taken care of first. Luckily, the machine automatically adjusts the amperage for you when you set your electrode wire feed speed (WFS), so you really just need to make sure you have a handle on the feed speed and voltage settings.

Table 10-2 shows you how to make the appropriate settings on your short circuit transfer machine for different types of shielding gas and electrode wire diameters on a steel mig welding project. (Flip to Chapter 9 to read up on the different types of transfer used in mig welding.) Some of the ranges are rather wide, but that's because you can weld with a wide range of metals that affect your settings.

When carbon dioxide comes out of the bottle, it's –104 degrees Fahrenheit, which, as you may imagine, cools off the weld pool. That's why using that gas for your mig welding adds one to two volts to your voltage. I've already accounted for those added volts in the Table 10-2 listings.

Table 10-2 Settings for Mig Welding Steel (Short Circuit Transfer)

Shielding Gas	*Electrode Diameter (in in.)*	*Amperage Range (in amps)*	*Wire Feed Speed Range (in in./minute)*	*Arc Voltage Range (in volts)*
75% argon/25% CO_2	.023	20 to 90	100 to 450	14 to 19
100% CO_2 (Adds 1 to 2 volts)	.030	30 to 145	90 to 340	15 to 21
100% CO_2 (Adds 1 to 2 volts)	.035	40 to 180	80 to 380	16 to 22
100% CO_2 (Adds 1 to 2 volts)	.045	75 to 250	70 to 275	17 to 25

Table 10-2 works for any welding position except vertical-up welding and vertical-down welding (both of which I cover in "Making vertical mig welds" later in the chapter). For vertical-up welding, subtract 30 percent from all the settings in the table. For vertical-down, add 30 percent to the settings.

If you're going to be using spray arc transfer rather than short circuit transfer, you need to set your machine a little differently. The details are in Table 10-3.

You must use at least 80 percent argon in your shielding gas in order to use spray arc transfer when you're mig welding.

Table 10-3 Settings for Mig Welding Steel (Spray Arc Transfer)

Shielding Gas	*Electrode Diameter (in in.)*	*Amperage Range (in amps)*	*Wire feed speed range (in in./minute)*	*Arc Voltage Range (in volts)*
Argon + 1 to 5% oxygen	.030	130 to 235	335 to 650	24 to 28
Argon + 1 to 5% oxygen	.035	165 to 310	340 to 630	24 to 30
Argon + 1 to 5% oxygen	.045	200 to 375	225 to 400	24 to 32
Argon + 1 to 5% oxygen	1⁄16	280 to 500	200 to 300	26 to 34

Trying Out Mig Welding

After you have your equipment set up and you know what kind of electrode wire feed speed and voltage settings you need for your mig machine (see the preceding sections), you're ready to try out some mig welding. No need to weld two pieces together yet — you can just practice making mig weld beads on a steel plate with a thickness of about 1⁄4 inch. Just lay it down flat on your welding table for this first practice exercise. (The plate doesn't need to be anything beautiful because you're just going to be practicing on it.)

Use an ER70S-2 electrode wire with a thickness of .035 inch. Start by setting your machine with a wire feed speed of 375 inches per minute and a voltage of 25. Set your shielding gas at 22 cubic feet per hour and follow these steps.

1. **Make sure you're wearing a welding helmet that protects your eyes from the rays of the arc.**

   Use a shade between #8 and #13. Need more info on eye protection? Check out Chapter 3.

2. **Hold the mig welding gun at a comfortable angle so that the contact tip is about 1⁄2 inch away from the steel plate.**

3. **Pull the trigger on the mig gun to start the arc.**

4. **Move the gun slowly left to right across your body, like you're writing on a tablet, paying attention to the weld pool size, metal transfer, and how much spatter (see Figure 10-1) you're producing as you lay down the weld bead.**

For best results, keep your electrode wire on the leading edge of the weld pool. If you weld on top of the pool, you don't get adequate penetration.

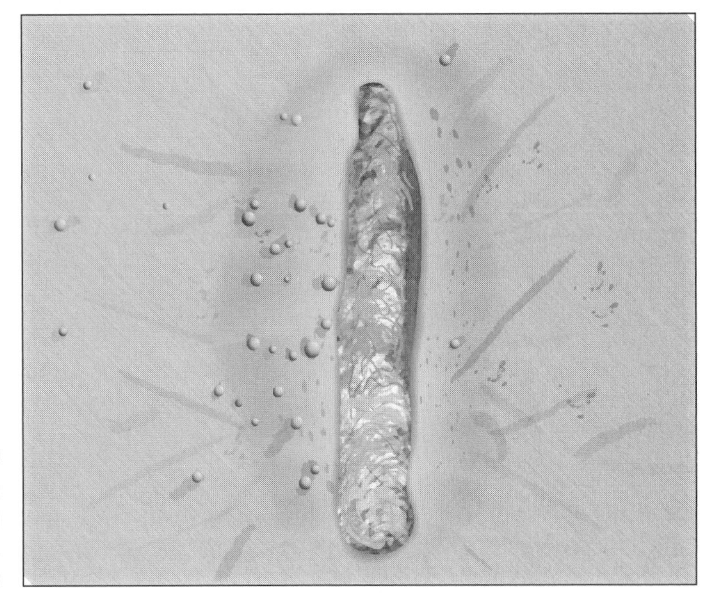

Figure 10-1: Typical spatter.

Look at the bead you've welded. Is it acceptable? You can see examples of good and bad mig weld beads in Figure 10-1. If so, you can turn off your mig machine and your shielding gas flow and then clean up any mess you've created. If it looks more like the poor weld in Figure 10-2, just keep trying until you get a good weld.

When you have that first mig welding bead under your belt, I think seeing how slight changes to your settings affect the quality of your weld is a useful step. Tables 10-2 and 10-3 earlier in the chapter can help you with that to some extent, but you also want to begin to get a feel for these things without consulting a chart.

As you're making these adjustments, pay attention to effects like the sound that the welding process makes and the amount of spatter you create. If your weld sounds like frying bacon or creates excessive amounts of spatter, or if your electrode wire *stubs off* into the weld pool (the wire hits the steel before it welds), you haven't found the right settings yet. If any of these results occur, always start by fine-tuning your voltage, because that's often the culprit.

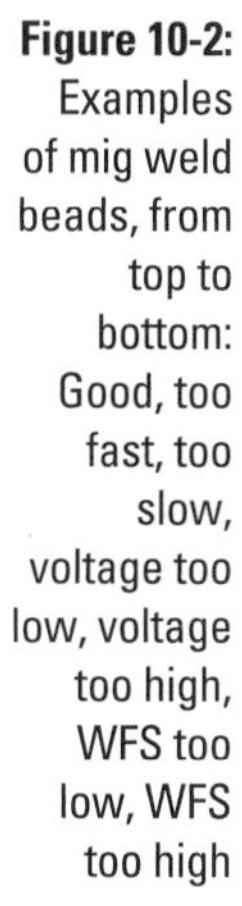

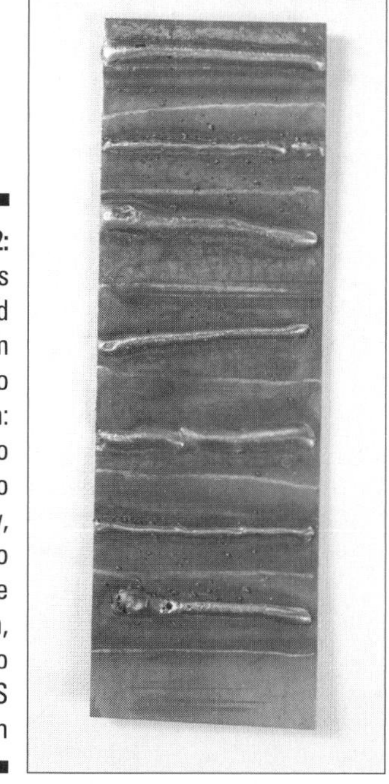

Figure 10-2: Examples of mig weld beads, from top to bottom: Good, too fast, too slow, voltage too low, voltage too high, WFS too low, WFS too high

Figuring out how to tweak your settings to give you the best possible weld is a skill that comes with practice. Using the same equipment as the preceding exercise, turn on the mig machine and shielding gas flow and double-check again to make sure you have all your protective gear on and in the right place. Then use the following steps to experiment with your settings.

1. **Set your wire feed speed and voltage about one-third of the way from one side of your dial (at 3 or so for the feed speed and about 18 for the voltage).**

2. **Give your gun ⅜ inch of stickout by pulling the trigger so that the wire feeds out past the contact tip.**

   To accomplish this, you can stop it when ⅜ inch is exposed, or run it out longer, let go of the trigger, and cut it back with a wire cutter.

3. **Pull the mig gun trigger and weld a 1-inch bead.**

4. **Now give your gun ¾ inch of stickout and weld another 1-inch bead.**

   Notice that the amount of spatter produced is greater when your stickout is longer.

5. **Repeat Step 3 with the wire feed speed and voltage set to the halfway point of your dial and ⅜ inch of stickout.**

   How did it go? Which weld looks best — this one or the other two beads you welded earlier with different settings? That best bead indicates the best settings for your welding machine. Also consider whether the sound changed or the wire seemed like it was forcing its way into the weld puddle as you played with the settings. Paying attention to these effects helps you fine-tune your setting-tweaking skills.

Making vertical mig welds

In a perfect world, you'd be able to do all your mig welds horizontally on a flat surface, which is widely accepted as being the easiest position to work in. Unfortunately, of course, the world isn't perfect, and you're probably going to need to be able to mig weld in the vertical position from time to time. Here's how you can mig weld vertically, going *uphill* (from bottom to top):

1. **Put on your safety gear and set your wire feed speed at 320 inches per minute and your voltage at 17 volts.**
2. **Tack weld your steel plate onto the angle iron of your welding table so it's standing in a vertical position.**

   A *tack weld* is a small weld used to hold parts or pieces in place until you make the finish weld.

3. **Hold the welding gun at 5 to 15 degrees to the direction you're going to be welding.**
4. **Pull the trigger on the gun and start welding at the bottom of the plate, using a slight side-to-side weave pattern as you work your way up the plate.**

   Does the weld sound right? If not, stop and adjust your voltage. Your weld pool should be small and under control. If it's too big to control properly, your voltage is too high. As you work your way up the plate, the weld pool should cool, forming a shelf that can support the next weld bead you weave across the plate.

You can also make vertical mig welds *downhill,* or from the top to the bottom of the piece. The process is pretty similar to the uphill procedure; you just set your wire feed speed to 600 inches per minute and your voltage to 26 volts. You can use the same side-to-side weaving pattern from the preceding exercise if you want, but you don't have to.

Joining pieces of sheet metal

The mig technique works wonderfully on sheet metal, so in this section I show you how you can easily join two pieces of sheet metal through mig welding.

For this sheet metal mig welding practice exercise, you need two pieces of 2-inch-x-8-inch 16 gauge sheet metal and an ER70S-2 electrode with a .035-inch diameter. Here are the proper equipment settings:

- Wire feed speed of about 350 inches per minute. (Amperage for that wire feed speed should be about 130 amps.)
- Voltage level of 18 volts.
- Stickout of 3/8 inch.
- Shielding gas made up of a 75 percent argon 25 percent oxygen mix, with a flow rate of 30 cubic feet per hour.

A good joint for practicing is the *lap joint* (a joint made by fusing a piece of metal lying on top of another piece). Here's how you can do it.

1. **Lap one piece of the sheet metal about 1/2 inch over the other piece flat on your welding table and tack them together with a very small tack weld.**
2. **Start at one end of the lap joint and weld a bead along the joint.**

   Keep a close eye on the molten pool to ensure that you're getting adequate fusion and also that the width of the weld doesn't get out of control.

   To help keep your bead straight, take your time and try to maintain a steady hand.
3. **When you reach the end of the joint, immediately go back to the start of the joint and weld another bead.**

   You may want to repeat this step one or two more times. Your weld is sufficient when the groove is full.

Watching Out for Common Mig Welding Defects

In Chapter 21, I tell you all about some defects that can wreak havoc on any type of weld, but if you're reading this chapter, right now you're probably concerned with making good *mig* welds. Never fear; this section lets you in on the defects that are more likely to happen in your mig welding work. Here are some problems to watch out for:

- **Incomplete penetration:** If the settings on your mig welding machine are too low, you can end up with a low level of penetration, which results in weak, lousy welds. To fix the problem, increase your wire feed speed and reduce your electrode wire stickout.
- **Excessive penetration:** This problem happens when you have too much heat in the weld area. Correct it by dialing down your wire feed speed and increasing your *speed of travel* (how quickly you weld the bead). You can also experience excessive penetration if the joint you're trying to weld is too big; if that's the case, just reduce the size of the gap.
- **Whiskers:** *Whiskers* are short lengths of electrode wire that stick through the finished weld on the back side of the joint. They're caused when the wire goes past the leading edge of the weld pool and melts off. You can prevent whiskers by slowing down your speed of travel and adding a slight side-to-side weaving motion to your mig welding. Just don't let your electrode wire get ahead of your weld pool, and you should be fine.
- **Incomplete fusion:** This defect is also called *overlap,* and it's caused by a slow speed of travel and inadequate heat. To prevent it from happening, keep your electrode wire on the leading end of your weld puddle, make the puddle smaller, and work with a shorter stickout.

Part IV

Getting Fancy: Plasma Cutting, Oxyfuel Cutting, and Other Processes

"My gosh, Barbara, if you don't think it's worth going a couple of weeks without dinner so we can afford a bottle carrier for the welding machine, just say so."

In this part . . .

Most of the welding done today — both hobby welding and commercial welding — uses arc welding equipment. But arc welding isn't the only game in town; the world of welding covers a wide range of processes, and many of them don't fit neatly (or at all, for that matter) into the arc welding category.

For example, welding with a torch that runs on oxyacetylene gas is an option. You can also practice soldering or braze welding (also called brazing), two more techniques that fall in the "special processes" category. Finally, a whole variety of cutting processes aren't welding — after all, you're cutting metals apart instead of welding them together — but are still really useful to know if you're going to be taking on any substantial welding projects. This part takes you inside those methods and processes.

Chapter 11

Examining Plasma and Oxyfuel Cutting

In This Chapter

- Discovering plasma arc cutting and equipment
- Making sense of oxyfuel cutting and its tools

Welding is the quickest, most efficient way to join metals. But what if you need to do the opposite? Instead of welding two metals together, you may need to be able to cut metal effectively. So how do you do it?

You can always use a saw, but if you've ever tried to cut metal with a saw you know it can be difficult, imprecise, and time consuming. Thankfully, two terrific options for cutting metal are straightforward enough that beginning welders can take advantage of them. These two processes are plasma arc cutting and oxyfuel cutting, and they're the focus of this chapter.

Read on to get the scoop on these two great cutting methods. I provide you with some good general info, present the advantages and disadvantages of both, and fill you in on the equipment you need to give plasma arc cutting or oxyfuel cutting a try.

Understanding Plasma Arc Cutting

You may remember the three states of matter (solids, liquids, and gases) from science class back in your school days. What many people don't know (or forget) is another state of matter called plasma. You can relax — I'm not going to go into a physics or chemistry lecture here. For the purposes of this chapter, just know that *plasma* is ionized gas, and you can make a material reach the plasma state of matter by heating up its gaseous state. If you want to get the kind of plasma that's needed to cut metal, you have to apply quite a lot of heat to a gas.

Plasma can cut through metal like a hot knife through butter. Here's how plasma arc cutting takes advantage of the properties of plasma to cut metal: Plasma cutting machines (I cover those a little later in this chapter) shoot pressurized gas over an electrode. The machine runs massive amounts of electricity through that electrode, and when you get it close enough to a piece of metal, a circuit is formed. That creates a huge amount of heat (about 30,000 degrees Fahrenheit), which turns the gas into plasma just before it reaches the metal.

Plasma cutting produces very high temperatures and very bright, intense light. Make sure you have all your safety bases covered before trying this process. (Flip to Chapter 3 for more information on welding safety.)

Identifying some good materials for plasma cutting

Plasma is so hot that it can cut many different kinds of metals, but plasma cutting works best with the metals that are good at conducting electricity. Some of the metals that you can very easily cut by using plasma cutting include

- Aluminum
- Magnesium
- Stainless steel
- Brass
- Copper
- Cast iron
- Mild steel
- High nickel steel
- Most other *non-ferrous* (iron-free) metals

You can use plasma cutting in all positions, but if you're using it in any position other than flat, you need to make sure you use only the low-power plasma machines that crank out less than 100 amps. The high-power machines are too dangerous to use out of position.

Taking a look at plasma cutting's advantages and disadvantages

As with all welding processes (and really, most things in life), plasma cutting has its pros and cons. The following list lets you in on the benefits that come with using plasma cutting.

- **Plasma cutting machines don't require any gas cylinders.**
- **You can cut a huge range of metals.**
- **Plasma cutting is fast.** You can cut four times faster with plasma cutting than you can with oxyfuel cutting when you're working with metals less than ½ inch thick.
- **You can plasma cut in all positions.**
- **You can get slag-free cuts on carbon steel, stainless steel, aluminum, and more.** *Slag* is waste material that attaches itself to the bottom side of the metal you've just cut.
- **The cuts are extremely precise, and the kerf is very narrow.** *Kerf* is the unintended loss of material on either side of a cut.
- **You can cut metals with little to no risk of *distorting* them (changing their shapes).**

Plasma cutting is an amazing process, but it's not perfect. Following are some of plasma cutting's downsides.

- **It's not cost-effective for cutting thicker pieces of steel because other cutting processes are faster (time is money, after all!) and use less energy.**
- **It requires electricity, so it's not as portable as oxyfuel cutting.** You always have to be close to a source of electricity for plasma cutting, which isn't the case with oxyfuel cutting. (I discuss oxyfuel later in the chapter.)
- **Most plasma cutters require a clean source of air in your shop.** Using shop air for plasma cutting makes your shop dirty.
- **Plasma cutting presents a risk of serious electrical shock.**
- **Your plasma cuts will always have a slight bevel (about 7 degrees) because of the way the machines cut.** That's less of an issue when you're working with thin materials, but if you're cutting thick metals, you may have to grind that bevel out of them.

Perusing and Preparing Plasma Arc Cutting Equipment

When I tell people about plasma arc cutting, they often assume it involves ridiculously expensive, hard-to-operate space-age equipment. Those folks are always surprised when I tell them that the equipment and tools you need for plasma cutting look like they belong in a welding shop (not on a space shuttle), the prices aren't any higher than what you'd spend on other important items in your welding shop, and you don't have to get a PhD in plasma physics to operate the machines. In the following sections, I provide you with details about plasma cutting equipment and what you have to do to get it set up.

Getting a handle on plasma cutting equipment

As I note earlier in the chapter, plasma arc cutting machines use a circuit to heat a gas to its plasma state. Figure 11-1 shows a diagram of what happens at the site of a plasma cut.

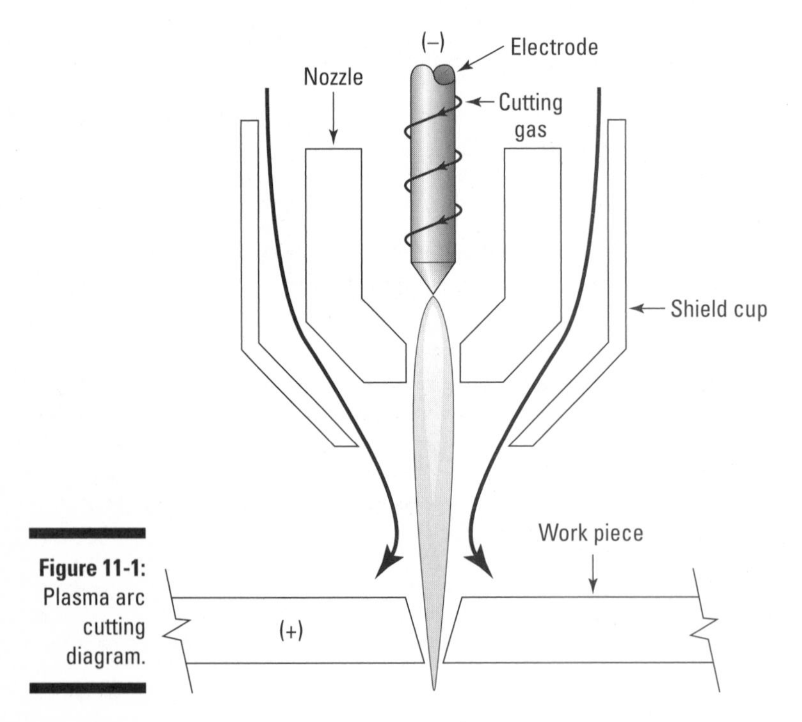

Figure 11-1: Plasma arc cutting diagram.

You may think that a piece of equipment that can do all that would look like something out of a science fiction movie, but plasma arc cutters are a lot smaller than you'd think (usually a box about two feet cubed), and at a quick glance they can look like a run-of-the-mill stick welding machine or similar piece of equipment. Check out Figure 11-2 for an example. They include a portable cabinet that houses all the electrical components, as well as an air valve assembly for hooking up the pressurized gas. They also have a *regulator assembly,* which uses a pressure regulator to keep the gas at the right number of pounds per square inch.

Figure 11-2: A plasma arc cutting machine.

A handheld plasma cutting torch connects to the main unit through a cable with three leads. (Kind of amazing that you can hold something in your hand that can create plasma with a temperature of more than 30,000 degrees Fahrenheit, isn't it?) Smaller torches are air-cooled, but the larger, higher-powered torches use a water-cooling system. The torch also includes a nozzle that protects the internal parts.

Here are a few other components of the plasma cutting torch that you should be familiar with because they wear out quite easily and need replacing every once in a while:

- **Electrode tip:** This piece is usually made of copper, with a small piece of tungsten attached.
- **Nozzle insulator:** This item is located between the nozzle tip and the electrode tip. The space between these parts is called the *electrical setback*, and it's critical to the proper operation of the plasma cutter.
- **Nozzle tip:** The *nozzle tip's* close-fitting parts restrict the air in the presence of electricity so the plasma can be generated. The small hole in the center of the nozzle tip changes and wears down over time, which changes the plasma action and results in below standard cuts if you don't replace it.

Setting the equipment up properly

Believe it or not, setting up most plasma equipment is a simple process. That's kind of nice because if someone calls to ask what you're doing, you can say, "Oh, well, not much. Just taking a few minutes to set up a plasma arc cutting machine." They'll think you're doing something really complicated, and of course you don't have to tell them otherwise unless you want to.

Here's how you get a plasma cutting machine ready to use.

Don't operate any equipment (including plasma cutting machines) until you've read the manufacturer's owner's manual and instructions.

1. **Make sure all the connections are tight.**
2. **Visually inspect all the gas lines for leaks.**
3. **Turn on the air compressor and wait for it to get up to the required pressure for your plasma machine.**
4. **Turn on the power switch.**

Yes, that's it! If you want to know how to make cuts with a plasma cutting machine, head to Chapter 12.

Exploring Oxyfuel Cutting Basics

Like plasma arc cutting (which I cover earlier in the chapter), oxyfuel cutting is a way to cut metals by using heat rather than mechanical means (a saw, for example). You don't need to know every last detail of how the process works, but put simply, a flame from a fuel gas heats steel to its *ignition temperature,* at which point the steel oxidizes but doesn't melt. Then a high-pressure stream of pure oxygen is directed at the heated steel, which dramatically increases the rate of oxidation and also blows away the oxidized slag, cutting down through the steel. It's not quite as high-tech as plasma arc cutting, but on steel, it definitely gets the job done.

Oxyfuel cutting requires a fuel gas. The most commonly used fuel gases are acetylene, propane, natural gas, and oxymapp gas (a liquefied petroleum gas). Acetylene is extremely popular, so that's what I focus on in this chapter.

Considering what you can (and can't) cut with oxyfuel

Unlike plasma arc cutting, oxyfuel cutting works only on *ferrous* (iron-containing) metals. You can't oxyfuel cut most nonferrous metals and stainless steels because oxyfuel cutting requires that the metal oxidize rapidly, and those metals don't fit that bill. However, ferrous metals encompass a pretty wide group of metals, so rather than go through all of those, the following list details the metals you *can't* cut with oxyfuel cutting.

- Brass
- Cast iron
- Copper
- Stainless steel
- Aluminum
- High-nickel steel
- Most nonferrous metals

Looking at the pros and cons of oxyfuel cutting

Like any welding or cutting process, oxyfuel has its pluses and minuses. First, the benefits:

- The equipment is versatile because you can cut *and* weld with an oxyfuel torch.
- It doesn't require electricity, so it's more portable than plasma arc cutting.
- If you know what you're doing, a good oxyfuel cut can be straight and square, and it requires hardly any cleanup.
- It doesn't create as much shop-polluting dust or fumes as plasma cutting.
- You can cut steel up to 11 inches thick with a handheld oxyfuel torch. That's some seriously thick steel.

And the downsides:

- The cylinders of gas required can be heavy and cumbersome.
- You're more likely to see distortion in the metal you're cutting than what you can expect with plasma arc cutting.
- You can't cut as many different kinds of metals as you can with plasma arc cutting.
- It's slower than some other cutting processes.
- It's pretty easy to do, but it can be hard to do really well — making good-quality cuts requires some practice.

Checking Out and Setting Up Oxyfuel Cutting Equipment

Compared to plasma arc cutting, oxyfuel cutting equipment is extremely low-tech. But don't think of that as a drawback; in fact, the equipment's simplicity makes it versatile. The following sections give you the lowdown on what you need for oxyfuel cutting and how to set it up.

Examining oxyfuel cutting equipment

The setup for oxyfuel cutting is basically the same as what you need for oxyfuel welding, which you can read all about in Chapter 13. Here's a quick list of what you need, with an accompanying visual in Figure 11-3.

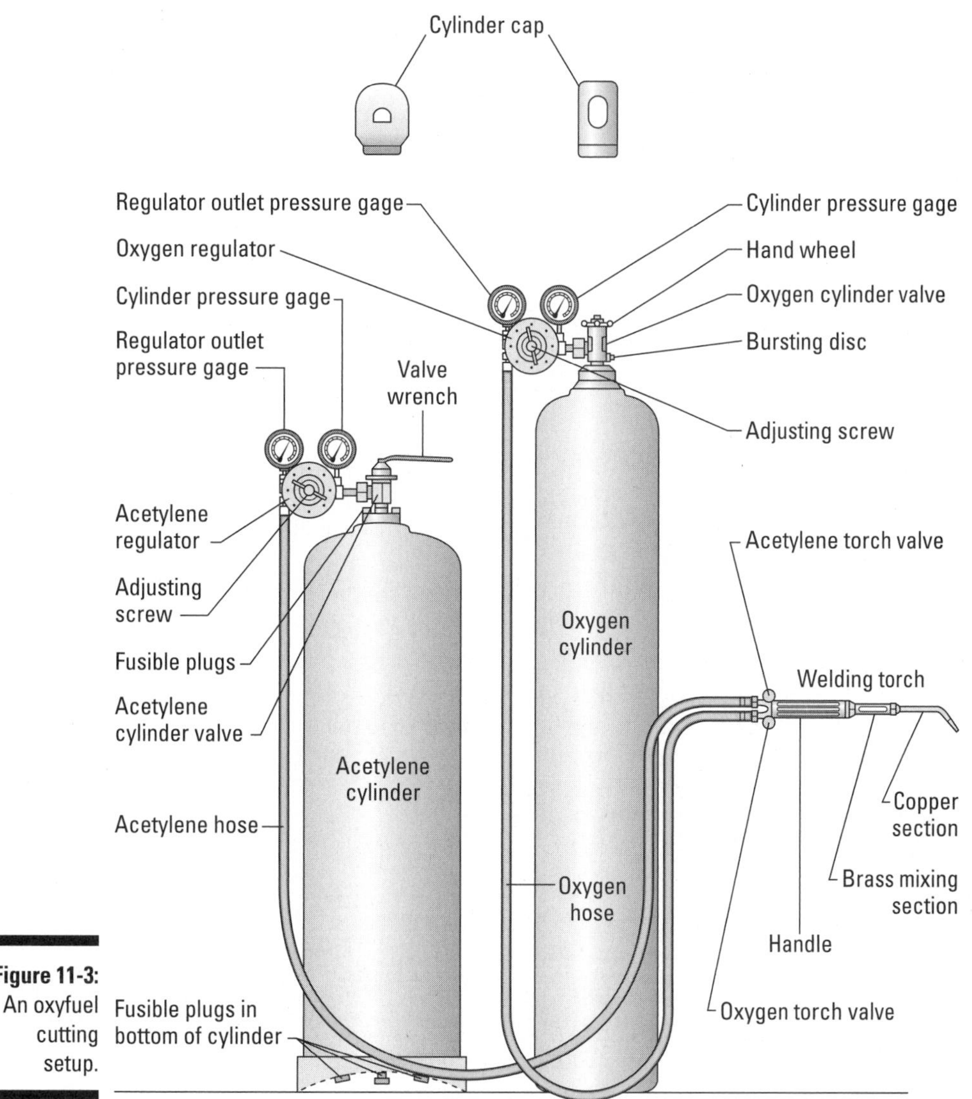

Figure 11-3: An oxyfuel cutting setup.

Don't use oil on any oxygen or acetylene equipment. Oil and oxygen react violently when they come in contact. If you need a lubricant for things like the threads on oxygen or fuel gas fittings, use water-based liquid soap (like the kind you probably use to wash your dishes in the sink).

As always, the most important equipment is your safety gear. Chapter 3 explains what you need to stay safe.

- **Gas cylinders:** You need both a cylinder of pressurized oxygen and a cylinder of pressurized fuel gas. Acetylene is probably the most common fuel gas, and it's the one I focus on in this section. (More on cylinders in "Deciding among different gas cylinder sizes" later in the chapter.) You can buy the cylinders at your welding supply store and return them when they're empty.
- **Hoses:** I'm not talking about the green garden hoses you use to water your plants. These items are specialty hoses designed to carry oxygen and fuel gas, and you can get them at your welding supply store.
- **Regulators:** These specialty valves control the amount of oxygen and fuel gas that flows into your cutting torch. You need one regulator for oxygen and one for fuel gas.

  Fuel gas fittings have left-handed threads, which means that the old "righty tighty, lefty loosey" saying doesn't apply. In fact, it's the opposite: You turn the fittings to the left in order to tighten them. The nuts used for fuel gas fittings have notches in them to help you identify which ones you need to turn left to tighten.
- **Friction lighter:** You use this item to light the oxygen and fuel gas mixture.
- **Cutting torch:** The torch is where the magic happens; it's the part of the equipment where the oxygen and fuel gas mix, and where the stream of oxygen that actually cuts the metal is controlled. Oxygen feeds from the cylinder through the hose and into the torch, where the flow is split so that some of the oxygen goes into a mixing chamber where it meets up with the fuel gas and the rest of the oxygen goes to the cutting jet. Check out the cutting torch in Figure 11-4.

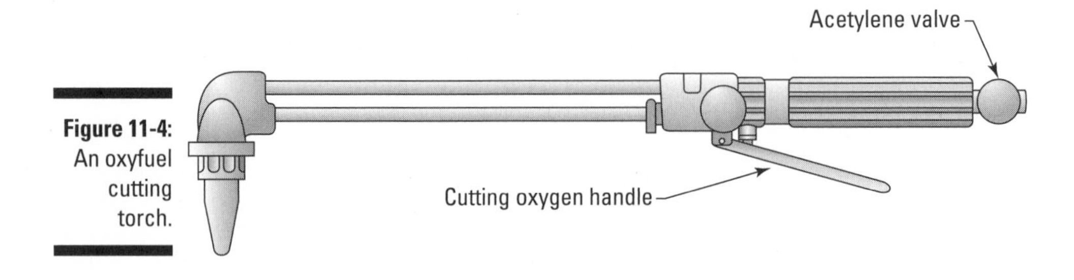

Figure 11-4: An oxyfuel cutting torch.

If you think you're going to want to use gas for both welding and cutting, consider getting a combination torch. They allow you to use the torch for both purposes.

- **Cutting tip:** The flame forms in this part of the torch. The tips are usually made of copper, but some are chrome. (Chrome tips are more expensive, but they last longer because *spatter* — small particles or balls of metal that stick to the tip during the cutting process — doesn't stick to the tip.) Cutting tips come in different sizes, and different cutting scenarios call for different tip sizes. Each size is assigned a tip number. Take a look at Table 11-1 to figure out which tip numbers match up with different metal thicknesses and gas pressures.

Table 11-1 Oxyfuel Cutting Tip

Tip Number	*Metal Thickness (in in.)*	*Acetylene Pressure (in psi)*	*Oxygen Pressure (in psi)*
0	¼	3	30
1	⅜	3	30
1	½	3	40
2	¾	3	40
2	1	3	50
3	1½	3	45
4	2	3	50
5	3	4	45
5	4	4	60
6	5	5	50
6	6	5	55
7	8	6	60
7	10	6	70

Deciding among different gas cylinder sizes

Oxygen and fuel gas cylinders come in different sizes. The size you choose depends on how much oxyfuel cutting (or welding) you plan to do, how often you want to have to change out the cylinders, and how portable you need your oxyfuel cutting setup to be. Here are the common cylinder sizes for the pressurized oxygen used in oxyfuel cutting.

- The big cylinders contain 244 cubic feet of oxygen. They weigh 150 pounds when full and 131 when empty.
- The medium-sized cylinders contain 122 cubic feet of oxygen. They weigh about 95 pounds when full and 83 when empty.
- The small cylinders contain 80 cubic feet of oxygen. They weigh 66 pounds when full and 62 when empty.

No matter the size of a pressurized oxygen cylinder, you can use them all until they're completely empty.

Here's a quick look at the two most common sizes of pressurized acetylene cylinders.

- The large size contains 300 cubic feet of acetylene (at a pressure of 220 pounds per square inch). They weigh 230 pounds when full and 215 when empty.
- The small size contains 100 cubic feet of acetylene (at a pressure of 225 pounds per square inch). They weigh 90 pounds when full and 84 pounds when empty.

You can also get pressurized acetylene in 40-cubic-foot and 10-cubic-foot cylinders, but these sizes aren't used very often, and you can't completely empty the smaller one; you have to stop using the gas when the bottle has a pressure of about 25 pounds per square inch.

Setting up oxyfuel cutting equipment

You can read all about how to make oxyfuel cuts in Chapter 12, but here's a rundown on how you should set up the equipment.

1. **Make sure the cylinders are placed together and secured either on a cart or against the wall so that you're sure they don't tip over or fall while you're cutting.**
2. **Remove the caps on the cylinders and keep them close by.**

   You'll have to put them back on when the cylinders are empty.
3. **Open and close the cylinders quickly.**

   That's called *cracking* the cylinders, and you need to do it to blow any dirt or other debris out of the valves.
4. **Attach the oxygen regulator to the oxygen cylinder with a crescent wrench or an open-end wrench; tighten the nut firmly and ensure the oxygen regulator adjusting screw is in the no-pressure position (turned counterclockwise until there's no resistance).**

5. **Open the oxygen cylinder valve very slowly; when you reach full pressure (about 2,000 pounds per square inch), open the valve all the way.**

   Don't ever stand directly in front of a gas pressure gauge because if it malfunctions, parts can blow through the front of the gauge and harm you.

6. **Connect the acetylene regulator, making sure that the connection nut is tightened firmly and that the adjusting screw is turned counterclockwise and has no tension.**

   Remember that the regulator has left-hand threads, so you need to turn it to the left to tighten.

7. **Slowly open the acetylene valve no more than 1½ turns.**

8. **Connect the red fuel hose to the gas regulator outlet and the green oxygen hose to the oxygen regulator outlet; connect the free ends of both hoses to your cutting torch and use an open-end wrench to snug them up.**

9. **Select the tip that fits the cutting job you're about to do.**

   Table 11-1 earlier in the chapter can help you make your selection. The tip is sealed with an O-ring, so you have to hand tighten it only — don't use a wrench.

10. **Slightly open the acetylene valve on the torch and adjust the acetylene regulator until the pressure corresponds to the tip you're using; close the valve on the acetylene side of the torch.**

   Always use the manufacturer's recommendations for proper oxygen and acetylene pressure. Sometimes you may have to turn up the oxygen pressure a little to cut through steel that has a coating of scale on it, but it's best to start with the manufacturer's recommendations and then simply turn up the pressure a little if you're having trouble making the cut.

11. **Repeat Step 10 for the oxygen valve.**

12. **Check your fittings for leaks by spritzing them with soapy water; if you see bubbles, tighten the fittings.**

Never use white thread seal tape on brass fittings in an effort to seal them. Brass fittings seal themselves.

Chapter 12

Ready, Set, Cut! Trying Out Plasma Arc Cutting and Oxyfuel Cutting

In This Chapter

- Giving plasma arc cutting a try
- Experimenting with oxyfuel cutting

Welding metals — that is, using massive amounts of heat to join them together — is a great thing, but sometimes what you really need is to cut metals apart. You may be surprised how often the need to cut metal pops up in your welding shop if you stick with welding for any long period of time.

You can use a saw to cut through small pieces of metal, but it's a pretty weak (and exhausting) option if your metal is big, thick, or heavy. If you're going to be doing a lot of metal cutting, consider exploring the options of plasma or oxyfuel. These two metal-cutting methods make the process easier and a whole lot faster. And to be honest, they're not that difficult to figure out after you understand a few of the basic principles and skills.

Read on in this chapter to get a feel for how you can take advantage of plasma arcs and oxyfuel to make short work of cutting metal in your welding shop.

Exploring Plasma Arc Cutting

Plasma arc cutting was designed to cut materials that couldn't be easily cut with oxyacetylene torches. As you can read in Chapter 11 (where I introduce plasma arc cutting), plasma arc cutting machines are so advanced, they make a complicated, remarkably high-energy process pretty easy to use. The following sections show you how you can use one of these machines to make different kinds of cuts in metal.

You can find dozens of different kinds of plasma arc cutting machines. They all work the same way, but some have more bells and whistles. Be sure you read all your machine's accompanying literature (instruction manuals and safety guides, for starters) before you even begin to make your first cut.

Like welding, plasma arc cutting can be dangerous if you don't take the necessary safety precautions before, during, and after the job. Flip to Chapter 3 if you need a reminder on how to stay safe while using welding or cutting equipment.

Here are a few things you need to get squared away before cutting metal with a plasma arc cutting machine.

- Make sure you have a welding helmet with a #5 shade, or a pair of safety goggles. Check out Chapter 3 for more on eye protection.
- Check your work area for any flammable materials. Because you're working with flame, you obviously don't want anything flammable anywhere near the area in which you'll be cutting.
- Turn on your air compressor and the plasma arc cutting machine's power switch.
- Set the regulator so you're getting air flow of at least 80 pounds per square inch.
- Press the trigger button to ensure air flow (note that the trigger button starts it all, including the pilot light).

Keep your feet out from under the area where you're cutting. If you don't, they can be hit with sparks, molten metal, or the piece of metal that you cut free when it falls to the floor.

To protect your cement floor from your welded piece, position a big piece of sheet metal, or a sheet-metal box full of sand, on the floor underneath the piece you're cutting. That way when the piece falls off, it will fall into the sand and cool off safely. If it hits the cement floor, it may cause the cement to explode.

Slicing a straight line

Straight cuts are an important part of welding because when you build projects, you need to cut pieces to a certain length, and those pieces are much easier to weld together if they have straight edges.

I recommend practicing a straight cut on an aluminum plate with a thickness of about 1⁄16 inch. Make sure the plate doesn't have any debris or coating anywhere on it. (You may need to do some grinding if the plate isn't clean.) To make a straight cut, just follow these steps.

1. **With a straightedge and a soapstone, mark a line on the metal piece to be cut.**

   Soapstones are tools largely composed of mineral talc that are used to mark metale.

2. **Place the piece on your work table so that the line you plan to cut is out over the edge of the table.**

   That way you don't cut into your table when you make the cut.

3. **Place a piece of angle iron along the cut line and secure both pieces to the table with a clamp, arranging the pieces so your clamp doesn't get in the way while you're cutting.**

4. **Attach a ground clamp to your piece or to the table to ensure that the piece you're working on gets the full amount of electricity during the welding process.**

5. **Press the torch trigger to start the pilot arc.**

6. **Use a continuous motion to move the torch along the line all the way down the cut.**

   Because you're cutting a piece of metal that's only about 1⁄16 inch thick, you can touch the torch tip to the work. If you were cutting something with a thickness of 1⁄8 inch or more, you'd want to maintain a space of about 1⁄8 inch between the torch and the metal piece.

7. **When you get to the end of the cut, release the trigger and hold onto the torch until the air stops flowing.**

8. **Turn off the power switch, clean up your tools, clean the weld area, and turn off the air compressor.**

You're done! If you made a good cut, the result should look like the straight cuts in Figure 12-1. If it doesn't, just keep practicing.

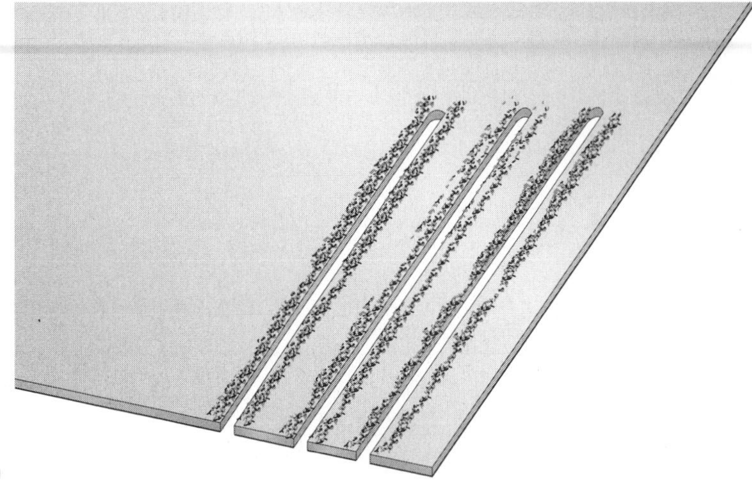

Figure 12-1: What straight plasma arc cuts in aluminum should look like.

Cutting a circle

Another important welding cut is the circular cut. Some welding projects require that a piece of metal (maybe a piece of pipe or tubing) pass through another piece of metal, and a circle-shaped hole is a great way to make that happen.

To cut a circle in a piece of aluminum with a plasma arc cutting machine, follow these steps.

1. **Using a soapstone, draw a circle on the metal piece that you want to cut.**

   If you want to practice on a perfect circle, trace around the outside edge of a washer or a jar lid.

2. **Clamp the metal piece to your work table so the circle hangs out over the edge of the table, making sure the clamp and table won't get in your way while you're making your cut.**

3. **Attach a ground clamp to your piece or to the table to ensure that the piece you're working on gets the full amount of electricity during the welding process.**

4. **Press the torch trigger to turn on the pilot arc.**

   When the arc comes on, you can immediately start cutting. If you plan for the large piece of metal with the hole in it to be your finished product, start cutting in the middle of the hole and then cut your way to the line before cutting all the way around. (Cutting on the line can be difficult if you try to start right on top of it.) If your finished piece will be the round piece that you cut out, start cutting on the outside of the hole and then cut your way down to and around the line.

5. **When you get to the end of the cut, release the trigger and hold onto the torch until the air stops flowing.**

6. **Turn off the power switch, clean up your tools, clean the weld area, and turn off the air compressor.**

Did you manage to cut a perfect circle? You can see an example in Figure 12-2. If not, you can always keep practicing.

Figure 12-2: Perfect circle cut in aluminum.

Creating a bevel

To cut a bevel with plasma arc cutting, first obtain a piece of aluminum at least ¼ inch thick. The rest of your setup should be the same as the one I describe in the earlier "Slicing a straight line" section, except that you need to use a C-clamp to secure the metal piece to your table in a way that allows you to cut a bevel through it. (Be sure that you have the piece clamped so that you don't accidentally damage your table or the clamp itself.)

After you've taken care of that, follow these simple steps.

1. **Attach a ground clamp to your piece or to the table to ensure that the piece you're working on gets the full amount of electricity during the welding process.**

2. **Angle the torch to the desired bevel angle and touch the tip to the piece you're cutting.**

3. **Press the torch trigger to start the pilot arc.**

4. **Maintain the desired bevel angle and smoothly work your way down the length of the piece, using a continuous motion to help ensure a good, clean beveled edge.**
5. **After you complete the cut, let go of the trigger and hold onto the torch until the air stops flowing.**
6. **Turn off the machine's power, clean up, and shut down the air compressor.**

Check out a successfully executed beveled cut in Figure 12-3.

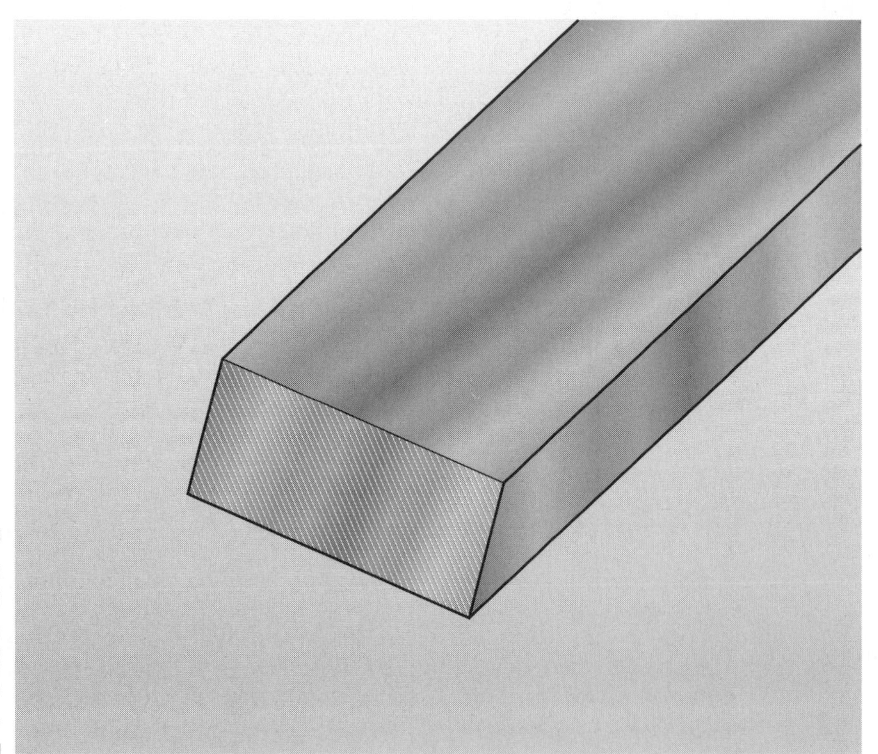

Figure 12-3: Example of a plasma arc bevel cut in aluminum.

Practicing Oxyfuel Cutting

In Chapter 11, I tell you all about the oxyfuel cutting process, and how useful it can be if you need to cut steel (or another *ferrous* — iron-containing — metal) quickly and efficiently. In the following sections, I walk you through the steps so you can actually practice making the cuts.

You need a welding helmet with a #5 shade for oxyfuel cutting. If you aren't yet familiar with welding helmets or the other important tenets of welding safety, take the time to read Chapter 3 before trying any kind of oxyfuel cutting. It can be a dangerous endeavor if you don't follow the right safety precautions!

To give oxyfuel cutting a try, make sure you have your equipment set up like I describe in Chapter 11. You can't expect to produce quality cuts if your setup is subpar or disorganized.

Never do the following actions when working with your oxyfuel setup:

- **Never disconnect the regulator from the cylinder without releasing the pressure out of the lines and making sure the cylinders are shut off.** If you don't shut off the valves, you release all the gas into your shop.
- **Never use any equipment (torches, regulators, cylinders, or hoses) that isn't in proper working condition.** If the equipment is compromised in any way, it isn't safe to use.
- **Never use tape to repair hoses used for welding or cutting.** Tape is a quick, shoddy fix at best, and you never know when it's going to fail on you.
- **Never use hose clamps on oxyacetylene hoses.** Hose clamps aren't a proven fix for oxyacetylene equipment. You can find *ferrules* (metal rings or caps) that are much better for that job.

Also, keep in mind that every hand cutting torch has a lever for opening and closing the cutting oxygen valve. When you're ready to start a cut, squeeze and press this lever slowly as you start the cut and then press it all the way down after you start the cut. Keep it pressed down until the cut is finished because you want to use a solid stream of oxygen throughout the whole cut, and pressing the lever all the way down provides just that.

Remove any flammable materials from your cutting area before you get started. You don't want to risk catching something on fire with sparks or hot slag.

Lighting the torch

When you've set the oxygen and acetylene pressures to the manufacturer's specifications for the size of the cutting nozzle you plan to use, you're ready to light your torch. The following steps show you how.

1. **Open the acetylene valve ⅛ of a turn and use a friction lighter to light the torch.**
2. **Open up the acetylene valve until the smoke clears and there's no gap between the tip and the flame.**
3. **Open the oxygen valve slowly until the two flames come together to form a *neutral flame*, which has a sharp inner cone.**

   Press the oxygen lever — the flame shouldn't change. If you press the lever and you get a feather of acetylene in the flame, shut the acetylene valve a little. If the flame instead changes to a blue hue, give yourself a little more acetylene.

Look at Figure 12-4 for an example of a lit oxyfuel torch.

Figure 12-4: Operating an oxyfuel cutting torch.

The torch flame may go out with a pop or a loud snap. This occurrence is a *backfire,* and it's caused by the tip touching the work or a droplet of hot metal. If this situation happens, shut off the torch, give it a moment, and then relight it. If another backfire happens soon after, your operating pressures are probably incorrect; check them and adjust accordingly. You may also have a loose or dirty tip or nozzle.

If the torch goes out and you hear a whistling or hissing sound, that's called a *flashback.* If you get a flashback, shut off your torch immediately, allow it to cool down, and check your operating pressures before you relight. After a flashback, I also recommend that you allow oxygen to flow through the torch to clean out any soot that may have accumulated before you try to relight.

After you make a cut and you're ready to shut off your torch, remember that you should always close the fuel gas first and then close the oxygen. If you're going to be stopping for more than an hour, shut off both cylinders and open both torch valves to release the pressure. Then back out the adjusting screws from both of the regulators and close the torch valves.

Making a straight cut

The following steps help you practice cutting a straight line by using oxyfuel cutting. I recommend practicing on a plate of steel about ½ inch thick and about 10 inches square.

1. **Use a soapstone to draw a straight line from one edge of the plate to the other.**
2. **Place the plate so it hangs off the edge of your welding table such that the line clears the table's edge.**

   You don't want to cut your table when you're cutting your plate!
3. **Put something under the plate and your table so that when you finish your cut, the free-hanging piece has a safe place to fall.**

   I recommend using a sheet-metal box filled with dry sand, or simply using another piece of sheet metal. This setup is smart because the floor in your welding area is probably made of cement, and cement can explode if some of these super hot materials come in contact with it.
4. **Make sure you have the appropriate tip on your torch for a ½-inch-thick piece of steel.**
5. **Put on your welding helmet.**
6. **Adjust the oxygen and acetylene to the levels recommended by your torch manufacturer.**

   Those numbers can vary from torch to torch, but if you stick with the manufacturer's recommendations, you should be in good shape.
7. **Light the torch, using Steps 1 through 3 from the preceding "Lighting the torch" section.**
8. **To start cutting, hold the torch at a 90-degree angle to the surface of the plate, with the inner cone of the flame about 1⁄16 inch away from the surface of the plate as shown in Figure 12-5.**
9. **Situate the nozzle so it's half on and half off the surface of the plate.**

   A red spot will appear slowly.

10. **Press the oxygen cutting lever so that the oxygen stream rapidly passes all the way through the plate.**
11. **Push the oxygen cutting lever all the way down and hold it.**
12. **Move slowly across the line you marked on the plate.**

    You can tell whether you're moving too slowly because the preheat flames will melt the top edges of the plate excessively (below the top of the surface). Conversely, if you move too fast, you'll lose the cut. If that happens, release the oxygen cutting lever, bring the steel back to a red heat, and slowly push back on the oxygen cutting lever to restart the cut.

When you've finished your first cut, the scrap section should fall down into the waiting sand or onto the piece of sheet metal you placed on the floor. If it doesn't, your piece may be held together by *dross*, which is a section of scrap material made by the torch puddle as it passes through the cut steel. You may be tempted to knock it off with your torch head, but that can damage your equipment. Instead, shut down your torch (closing off the fuel gas first and then the oxygen) and use a hammer to break off the scrap piece.

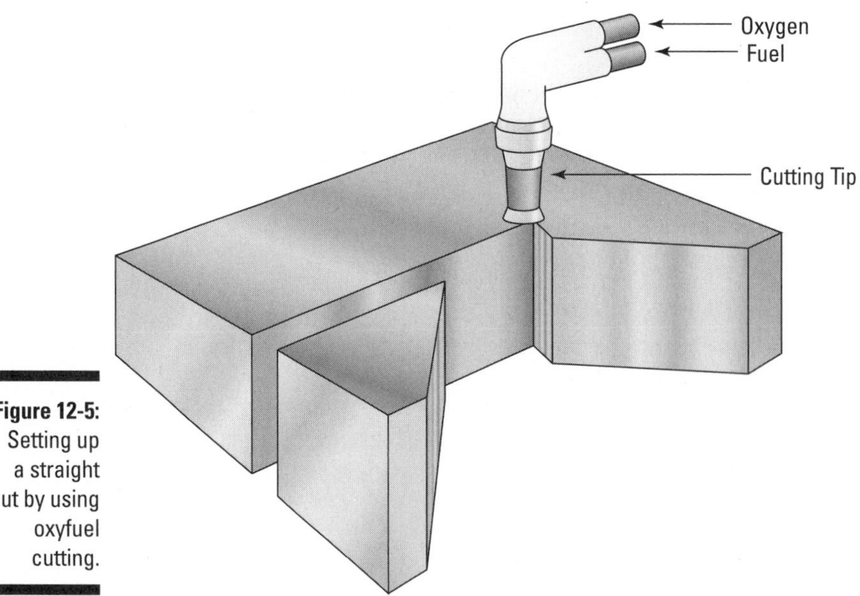

Figure 12-5: Setting up a straight cut by using oxyfuel cutting.

Cutting out a circle

If you practice oxyfuel cutting, you can experience the same benefits of being able to cut circles as plasma arc cutters do (see "Cutting a circle" earlier in the chapter). Here's how you do it, using a 10-inch-square piece of steel ½ inch thick.

1. **Draw a circle 4 inches in diameter on the plate with a soapstone.**
2. **Clamp the plate to your table so the circle hangs off the table's edge.**
3. **Place a piece of sheet metal or a sand-filled sheet-metal box on the floor to catch the scrap piece.**
4. **Check to make sure that you have the right tip on your torch and your welding helmet on, and that your oxygen and acetylene levels match up with the torch manufacturer's suggestions.**
5. **Light the torch using Steps 1 through 3 from "Lighting the torch" earlier in the chapter.**
6. **Position the flame over the center of the drawn circle, and hold it there until a spot on the surface starts to melt.**
7. **Raise the torch tip about ⅛ inch to ¼ inch off the plate, and slowly open the oxygen valve; as you open the valve, move the torch tip off the center of the circle so the *slag* (debris) and molten metal are blown to one side.**
8. **Continue to move the tip until the oxygen jet has passed all the way through the piece.**
9. **Cut out your circle, following the guidelines in Step 4 of "Cutting a circle" earlier in the chapter.**

   You can see an example of such a cut in Figure 12-6.

If you plan to use oxyfuel cutting to cut a lot of circles, consider buying a circle-cutting attachment for your torch. Ask about them at your local welding supply shop.

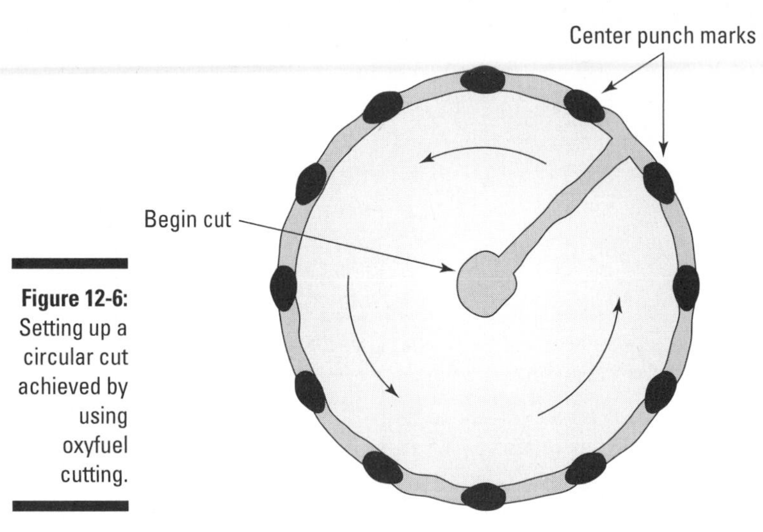

Figure 12-6: Setting up a circular cut achieved by using oxyfuel cutting.

Taking on a beveled edge

Cutting a beveled edge with oxyfuel cutting is a lot like completing a straight cut, except it's somewhat more difficult to master. You can practice a bevel with a similar (½-inch-thick, 10-inch-square) piece of steel, but you may want to go ahead and get a small stack of those plates to practice on because few welders get the bevel right on the first try. (Of course, I have all the faith in the world in you — I just know how hard cutting a bevel for the first time is.)

Much of the process for cutting a beveled edge is the same as the one for making a straight cut (see "Making a straight cut" earlier in the chapter), although you need to make sure from the get-go that you're using the right torch tip — the thickness you cut through actually increases because of the bevel. You can figure out tip size you need by simply measuring across the angle you're going to cut and then set your torch accordingly. For example, if you're cutting a 45-degree bevel, the thickness is 1.4 times the thickness of the plate. That would be 7⁄10 inch, but chances are you don't have a torch tip for that measurement. If you have a tip for 11⁄16 inch, that'll work.

Using oxyfuel to cut a beveled edge is tricky, so I recommend going through the motions with an unlit torch a few times to get used to how making those kinds of movements feels. Holding the same angle throughout the cut is difficult and may feel pretty odd at first.

When you're ready to practice with a lit torch, you can set up the work area in much the same way as you do to practice oxyfuel cutting a straight line. You just have to cut on an angle, and there's really no better way to get good at that skill than to start trying it out. See Figure 12-7 for the setup.

When you're practicing cutting a bevel, keep your cutting speed as steady as possible. And remember that you can't go as fast as you would if you were cutting a straight line because you're actually cutting through more material because of the angle.

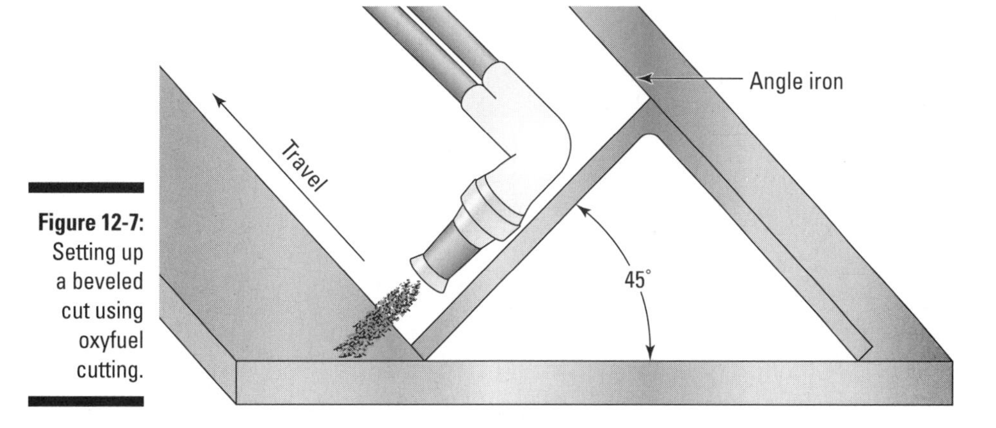

Figure 12-7: Setting up a beveled cut using oxyfuel cutting.

Chapter 13

Exploring Special Weld Processes

In This Chapter

- Taking a look at gas welding
- Singling out soldering

I devote quite a bit of this book to stick, tig, and mig welding, but those aren't the only welding processes you can use. Other options include gas welding, fusion welding, brazing, and soldering, and in this chapter, I fill you in on the basics of those techniques.

Working through the Basics of Welding with Gas

Gas welding involves joining metals by melting them with a gas flame. You can use a gas torch for many different processes in your shop, including welding, cutting, *annealing* (heating up steel and then letting it cool to make it less brittle), and hardening. For the purposes of this section, I focus on what you can do in the way of welding with a gas torch.

The most common type of gas used for welding steel is oxyacetylene gas, which is a combination of acetylene and oxygen. But these options aren't the only gases you can use for welding. You can also use oxymapp gas. *Oxymapp* is a liquefied petroleum gas, and it burns at about 5,300 degrees Fahrenheit, which is a lot lower than what you can get with oxyacetylene. That lower temperature means it takes longer to preheat and weld steel, and so oxymapp makes it a pretty weak choice for gas welding. However, you can use another gas, propane, for brazing and soldering — read more about those two techniques later in this chapter — and oxyfuel cutting, which you can check out in Chapters 11 and 12.

Many new welders wonder whether they can use propane for gas welding, but I don't recommend it — especially for working on steel — because the propane doesn't contain carbon, and that can cause your welds to be brittle.

Taking a gander at gas cylinders

In order for you to do oxyacetylene welding, you need a cylinder of pressurized oxygen and a cylinder of pressurized acetylene. These cylinders are made of steel, and they come in many sizes. The three most common oxygen cylinder sizes are 80 cubic feet, 122 cubic feet, and 244 cubic feet. Cylinders of that size are pressurized to 2,200 pounds per square inch (psi). The most common acetylene cylinder sizes are 60 cubic feet, 100 cubic feet, and 300 cubic feet, and they're pressurized to 220 pounds per square inch. Cylinders should come clearly marked as being oxygen or acetylene, but another way you can tell which is which is by the fittings. Oxygen cylinder fittings are right-handed, and acetylene cylinder fittings are left-handed. *Right-handed* fittings are probably what you're most used to; turn them clockwise to tighten them up. *Left-handed* fittings are the opposite — you turn them counterclockwise to tighten them. The left-handed fittings also have notches cut into the nuts to help you identify them.

All cylinders are equipped with a nifty safety device that releases the gases slowly if the temperature rises above the safety pressure of the cylinder. These devices are a must because gases expand when they're heated and contract when they're cooled. As a general precaution, though, be sure to keep your cylinders away from heat.

If you're planning to use oxyacetylene welding in a variety of areas, keep in mind that anytime you move the cylinders you need to be completely sure that the valves are turned off. If you need to move the cylinders around a lot, place them on a two-wheeled cart so they're easier to move. If you're oxyacetylene welding in your shop or another permanent work station, chain the cylinders to something sturdy so they can't be knocked over accidentally.

Never lay gas cylinders in the horizontal position; keep them upright at all times. And don't lift a gas cylinder by its valve protection cap.

If you need to move a cylinder a short distance, first make sure the cap is on tightly. Then rotate the cylinder on its bottom edge, put your hand on top of the cap, and tilt the cylinder toward you. Start the tank rolling by pushing it with your other hand, and roll it to wherever you need to go.

Leaking cylinders are very dangerous. If you find a leaking cylinder and you can't stop the leak by closing the valve, move the cylinder to an open air spot with nothing flammable or explosive within 150 feet. Then open the valve and slowly release the gas until the cylinder is empty.

Looking at more gas welding equipment

Gas welding requires some specialized gear in addition to gas cylinders. Here's what else you need for oxyacetylene welding:

- **Torch body:** You can get a good look at an oxyacetylene torch in Figure 13-1. Your oxyacetylene torch mixes the two gases and directs them to the torch tip, where the mixture is burned. I recommend getting an *equal pressure torch,* which supplies oxygen and acetylene at the same pressure. The two gases are mixed together in the mixing chamber, which is in the torch handle. (A few torches have the mixing chamber in the tip); you can find one at your welding supply store. Take care of your torch; when you're done welding, always hang it so it won't fall, and be sure the tip isn't touching anything that can burn or melt.

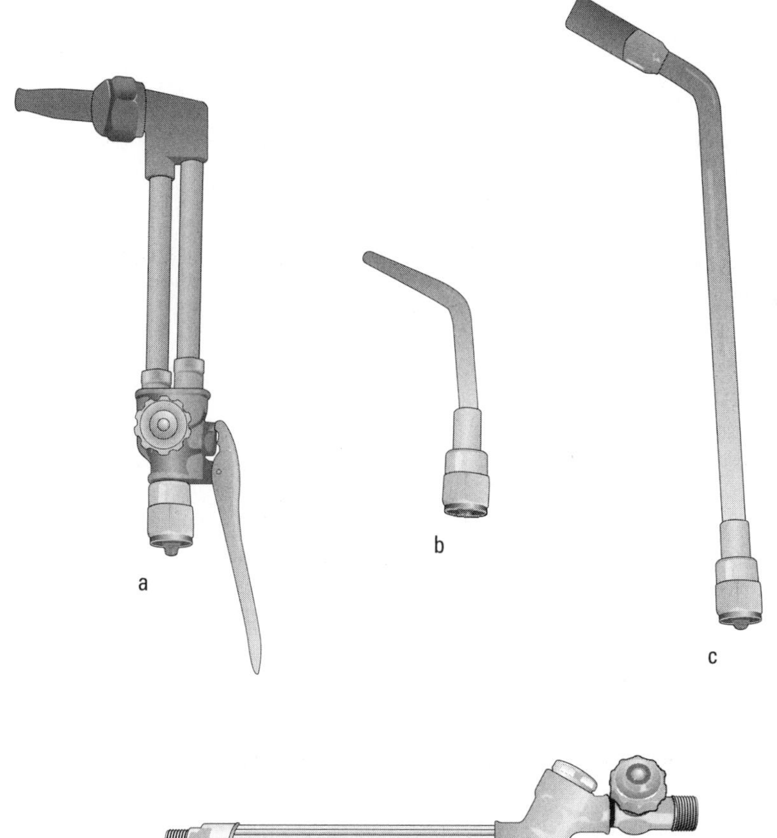

Figure 13-1: An oxyacetylene torch's cutting torch attachment (a), welding tip (b), heating tip (c), and torch body (d).

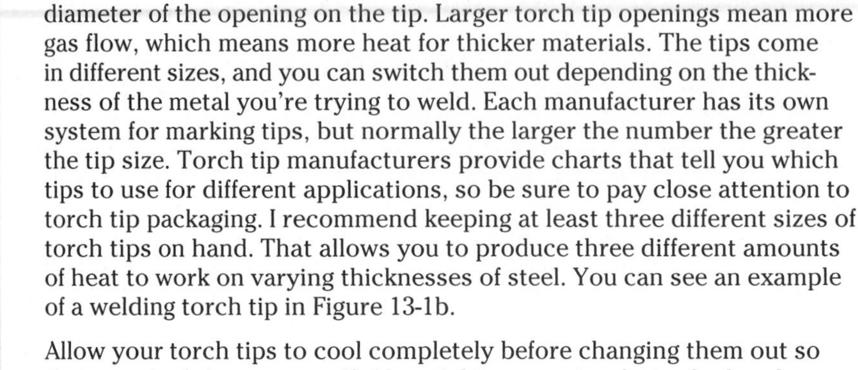

- **Torch tips:** *Torch tips* (where the flame forms) are sized according to the diameter of the opening on the tip. Larger torch tip openings mean more gas flow, which means more heat for thicker materials. The tips come in different sizes, and you can switch them out depending on the thickness of the metal you're trying to weld. Each manufacturer has its own system for marking tips, but normally the larger the number the greater the tip size. Torch tip manufacturers provide charts that tell you which tips to use for different applications, so be sure to pay close attention to torch tip packaging. I recommend keeping at least three different sizes of torch tips on hand. That allows you to produce three different amounts of heat to work on varying thicknesses of steel. You can see an example of a welding torch tip in Figure 13-1b.

Allow your torch tips to cool completely before changing them out so that you don't burn yourself. Also, tighten your torch tips by hand — don't use pliers — because the tips have O-rings that ensure a good fit without excessive force.

Molten metal droplets can get stuck in your torch tip openings. Use a tip cleaner (available at your welding supply store) in the correct size to unplug the openings.

- **Hoses:** You can get specialty hoses from your welding supply store that are made to carry oxygen and acetylene from your cylinders to torch. These hoses are usually joined together, with a green hose for your oxygen and a red hose for your acetylene. The length of the hose you need depends on where you're working, but I recommend getting hoses at least 25 feet long.
- **Regulators:** These specialty valves fit on your oxygen and acetylene cylinders and control the amount of gas that the cylinders release into the hoses. Regulators have gauges so that you can match up the flow of gas to the tip size you use. You need one regulator for oxygen and one for acetylene.
- **Goggles:** For gas welding, you need all the safety gear that I describe in Chapter 3, but you specifically need a welding helmet or goggles with a #5 shade for gas welding.
- **Friction lighter:** A *friction lighter* uses flint to create sparks, and it's the only type of lighter you should use for starting your gas welding flame. If you use matches or a cigarette lighter, you run the risk of burning your fingers.
- **Filler rods:** The *filler rods* (rods you put in the molten weld pool) you use for gas welding should be the same composition and thickness as the base metal you want to weld.

Never use oil or grease on any oxyacetylene equipment because oxygen and oil react violently with each other.

Getting to work with gas welding

After you have all of your gas welding gear assembled (see the preceding sections), you're ready to give the process a try. The following sections walk you through the process of getting the torch set up, lighting the flame, making the actual weld, and finishing up. I walk you through the process in the next few pages.

In this section, I give you general information on how to go about oxyacetylene welding. Your equipment may require a variation or two on these basic steps, so be sure you read all the equipment manuals and instructions before starting.

Gas welding is just as dangerous as any of the other welding processes if you do it recklessly or without taking the proper safety precautions. To familiarize yourself with the details of welding safety, read through Chapter 3.

Setting up the torch

Here's how you get your gas welding torch ready to use.

1. **Place the oxygen and acetylene cylinders together and secure them either on your cart or against the wall with a chain so that they can't fall down.**
2. **Remove the caps from the cylinders and put them in a safe place so you can put them back on when the cylinders are empty and ready to return to the welding supply store.**
3. **Open and close the valves on the cylinders quickly.**

   This process is called *cracking the cylinders,* and it's useful for blowing out any dirt or other debris that may be lodged in the valve.
4. **Attach the oxygen regulator to the oxygen cylinder by using a crescent wrench or an open ended wrench; tighten the nut firmly and ensure that the oxygen regulator adjusting screw is in the no pressure position.**

   You can verify the screw's position by turning the adjusting screw counterclockwise until you encounter no resistance.

   Never stand in front of a regulator gauge. If it malfunctions, the gas pressure can blow the pieces of the gauge into you.
5. **Open the oxygen cylinder valve very slowly to release a gradual amount of oxygen into the regulator; after you reach full pressure (approximately 2,200 pounds per square inch), open the valve all the way.**

   If you can't open a cylinder valve by hand, don't use a wrench. Return the cylinder to the supplier and exchange it for a different one.

6. **Connect the acetylene regulator; tighten the connection nut firmly and make sure the adjusting screw is turned counterclockwise until it has no tension.**

   Remember that the acetylene fittings have left-handed threads.

7. **Open the acetylene valve slowly no more than 1½ turns.**

   That amount may not seem like a lot, but keeping the turns to a minimum is important so that you can shut off the valve quickly if you encounter an emergency.

8. **Connect the red fuel hose to the outlet on the acetylene regulator and then the green hose to the outlet on the oxygen regulator; connect the free ends of the hoses to the torch and snug them up with an open ended wrench.**

9. **Select an appropriate torch tip.**

10. **Open the acetylene valve on the torch just a little and adjust the acetylene regulator until the pressure corresponds to the tip you're using; close the torch's acetylene valve.**

   Right now, you're just setting up the torch; head to the following section for information on how to start the flame.

11. **Repeat Step 10 for the oxygen valve.**

12. **Use a spray bottle full of soapy water to check for leaks and adjust your fittings as necessary.**

   In Chapter 3, I show you how to use soapy water to check for leaks.

Lighting the torch

Lighting an oxyacetylene torch is a pretty straightforward process, but you certainly want to be sure you get it right. Read through the following steps to find out how.

1. **Hold your friction lighter in one hand and your torch in the other.**

2. **Open the acetylene valve on the torch one quarter of a turn.**

3. **Point the torch away from you (and anyone else in the area); hold the friction lighter right next to the torch tip and strike it to light the torch.**

4. **After the torch lights, open the torch fuel valve until it stops smoking, making sure that the flame doesn't leave the tip.**

5. **Open the torch's oxygen valve until a bright inner cone appears on the flame.**

   When the feathery edge of the flame disappears and a sharp inner cone appears, you have a neutral flame. The type of flame you have depends on the amount of oxygen in the gas mixture. The *neutral flame* (see

Figure 13-2a) is the most commonly used flame because it doesn't burn the metals you work on. A neutral flame has a dark blue inner flame and a light blue outer flame. The two other kinds of flames:

- The *carburizing flame* (see Figure 13-2b) has an excess of acetylene relative to oxygen. You can easily spot a carburizing flame because it has a long, white, feathered area at the end of the flame. You can use a carburizing flame when you're working with silver, or for soft soldering (which I cover later in the chapter).
- The *oxidizing flame* (see Figure 13-2c) has an excess of oxygen relative to acetylene. It's lighter blue than a neutral flame. I suggest using an oxidizing flame when you're brazing brass. (More on brazing later in this chapter.)

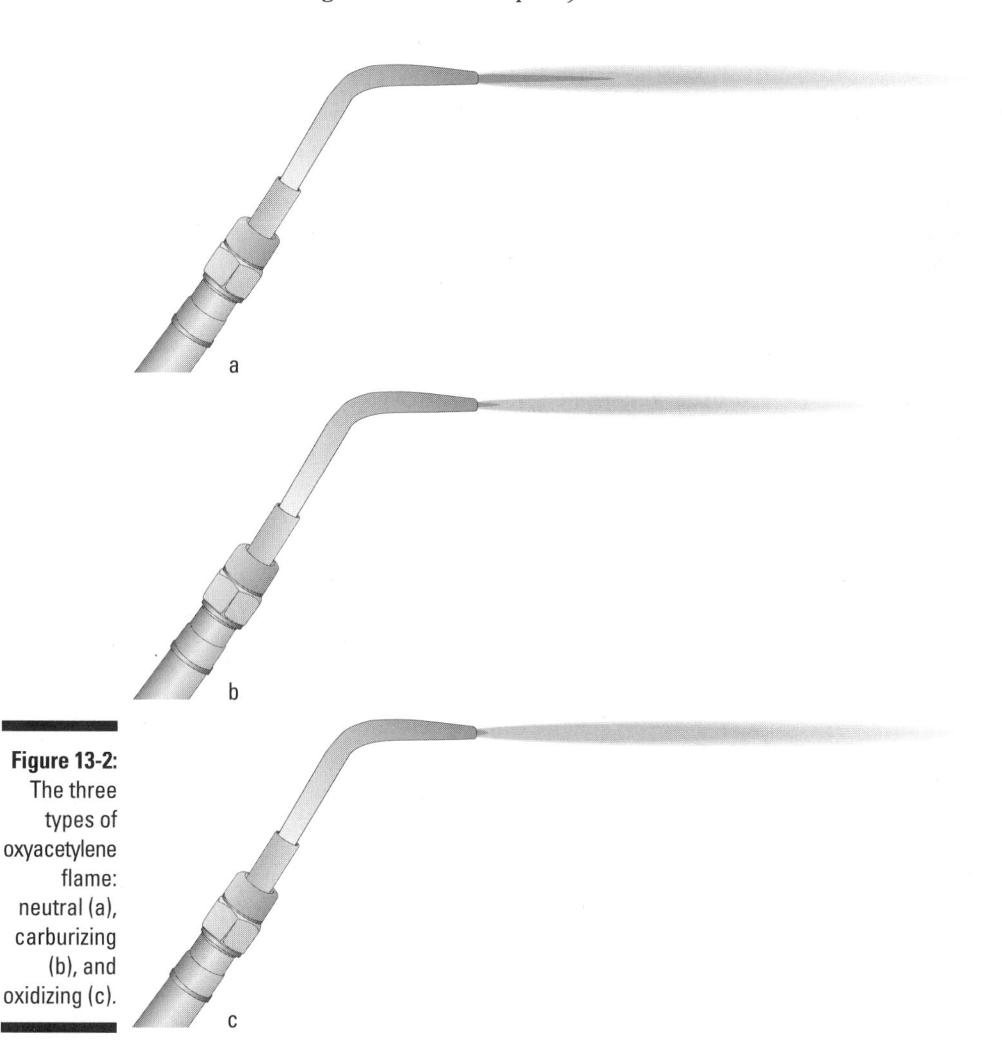

Figure 13-2: The three types of oxyacetylene flame: neutral (a), carburizing (b), and oxidizing (c).

Making the weld

The techniques you use and movements you make with the gas-welding equipment are very similar to those you use with tig welding (see Chapter 8). The biggest differences between the two processes are the heat sources and the metals you can weld with each process.

The most common method or position for gas welding is the forehand method, and I suggest you make it your go-to. It's the best method for gas welding materials that have a thickness of ⅛ inch or less. To use the forehand method, follow these steps:

1. **Grip the torch so it's pointed in the direction you're welding and hold it at a 45-degree angle.**
2. **Move the flame over the seam that you want to weld, moving it in small circles as you go.**
3. **Hold your filler rod at a 45-degree angle, slanting it away from the torch; as you continue moving the torch to create the weld puddle, dip the filler rod in and out of the puddle slowly so that it's continually heated up and melted.**

   Keep the puddle small and concentrate the flame in the puddle. That way, you preheat the base metal ahead of the weld (which is very helpful).

Finishing up

When you finish a weld (or when you're ready for lunch), you need to shut off the torch. Here's how to do it.

1. **Close the acetylene cylinder valve.**
2. **Open the torch's acetylene valve to release the pressure off the acetylene regulator.**
3. **Close the torch's acetylene valve.**
4. **Turn out the pressure adjusting screw on the acetylene regulator.**
5. **Repeat Steps 1 through 4 for the oxygen.**
6. **Give yourself a pat on the back.**

   Okay, this step is optional, but if you just successfully completed an oxyacetylene weld, why not give yourself a little credit?

Discovering Brazing (Braze Welding)

Brazing is a gas welding process in which you join metals by using heat that surpasses the 800-degree-Fahrenheit mark and a *nonferrous* (iron-free) filler rod with a melting temperature below that of the base materials. The most important point about brazing is that you can use it to join dissimilar metals — cast iron to steel, brass to steel, or copper to steel, just to name a few examples. The following sections give you the lowdown on brazing.

Keeping a few brazing rules in mind

A successful brazing job requires that you stick to a few rules, as follows:

- **The surfaces of the metals must be free of contaminants.** Use steel wool to clean off all the metals, and use *flux* (a material that dissolves or removes oxides and other contaminants from the surface during brazing) for additional cleaning when the surfaces are heated.
- **The joint to be brazed must have a tight fit.** You use two basic joints for brazing: the *butt joint* (two pieces of material lying together on the same plane) and the *lap joint* (two pieces of material overlapping each other, usually in a parallel plane). If the joint has sizeable gap, brazing just doesn't work. But if the joint is too tight, the melted filler rod can't penetrate the entire joint, and you get a weak, ineffective weld.
- **The base metals you're brazing must remain stationary during the heating and cooling process.** If the base metals move while you're welding or before everything has cooled off after welding, the joint's integrity will likely be compromised

Giving brazing a try

If you're trying brazing for the first time, I recommend going with an oxyacetylene torch in the flat position and using the forehand method I discuss in "Making the weld" earlier in the chapter. You can start by creating a brazed *corncob*, which is a piece created when you join two different metals, following the steps in this section.

Safety first! If you haven't read Chapter 3 (all about welding safety) and you don't have the proper safety equipment, don't even think about trying brazing. Pay special attention to ventilation; brazing filler rods contain zinc, and if you breathe the smoke that can come off those rods, you may end up with zinc poisoning.

1. **Acquiring a section of carbon steel pipe 1 inch in diameter and five inches long, as well as four ⅛-inch fluxed brazing filler rods that are 36 inches long.**
2. **Clean the pipe thoroughly with steel wool to remove any contaminants on the surface.**
3. **Lay the pipe between two fire bricks, leaving a ¾-inch space between the bricks.**

   Fire bricks are special bricks that can withstand extremely high temperatures; find them at your welding supply or hardware store.
4. **Set up and light the oxyacetylene torch, using the steps I describe in "Working through the Basics of Welding with Gas" earlier in this chapter, and adjust the torch so you have a neutral flame.**
5. **Preheat the pipe to burn off any grease or varnish that may be left on the surface.**
6. **If you're right-handed, start on the right end of the pipe and melt off a small portion of the filler rod onto the end of the pipe.**

   If you're left-handed, start on the left end of the pipe. The molten puddle should be very fluid. Be sure that when you use more of the filler rod, you let only the molten puddle (not the flame from the torch) melt the rod.

   If you notice white smoke coming from your molten puddle, that means you're burning the zinc out of your filler rod, which will result in a poor weld. Avoid that problem by welding at a lower temperature or moving more quickly with the molten puddle.

   When you're practicing the brazing process, you can *quench* the metal between passes with the torch. Use pliers to pick up the section of pipe you're practicing on and place it in a tank of water to cool it very quickly. Quenching isn't good for the integrity of the brazed weld, though, so don't quench unless you're just practicing.

7. **After you make your first pass with the torch, start a second pass by pointing the tip of the flame at the edge of the previous pass; when part of the first braze starts to melt, add some of your filler rod and proceed.**

   This pass laps over the first braze bead ⅓ to ⅕.

8. **When you've completed your second pass, quench the welded metal (or allow it to cool) and go ahead and start on a third pass.**

   When you're finished with the third pass, you should have a finished product that resembles the brazed corncob in Figure 13-3. For some brazing projects, you may need to make even more passes. (For example, projects involving thick pieces of metal definitely require more than three passes.)

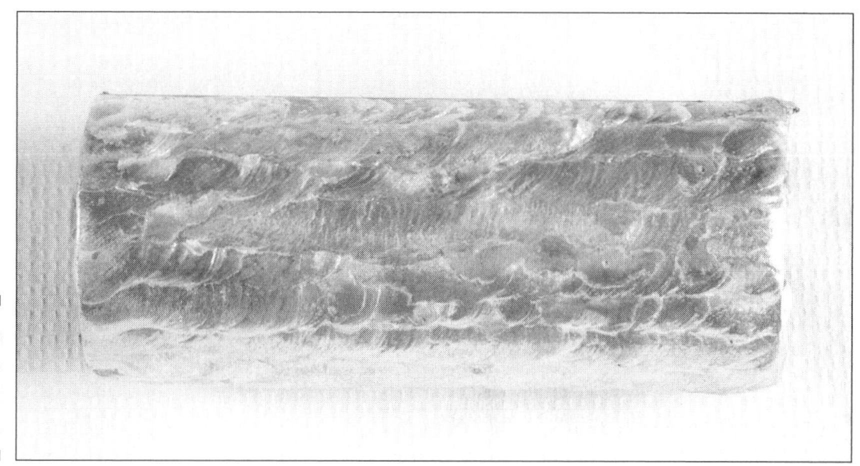

Figure 13-3: An example of a brazed corncob.

Finding Out about Fusion Welding

Fusion welding allows you to join two or more pieces of *ferrous* (iron-containing) metal together. You melt the ends or edges of the pieces with heat from a gas torch, and the molten portions of the metals flow together and form one. You can do fusion welding with or without a filler rod. To get a visual, take a look at Figure 13-4.

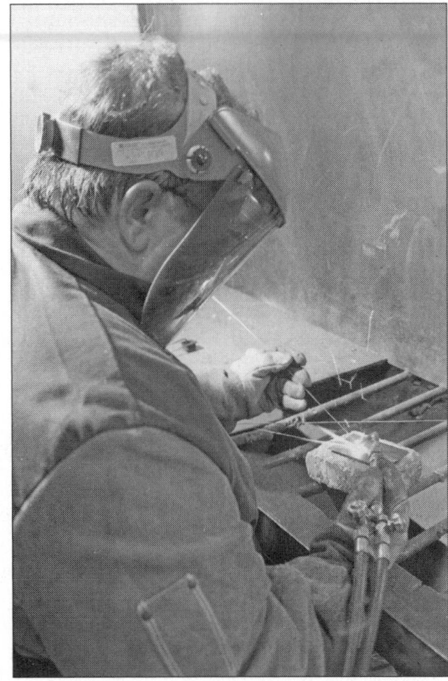

Figure 13-4: Fusion welding a butt joint with an oxyacetylene torch.

Getting set up for fusion welding is very similar to setting up for other types of gas welding, which I cover earlier in the chapter. You need to put the metal pieces next to each other in a way that allows them to melt, flow together, and form a joint while you're welding, so arrange your pieces accordingly before you get started. Then proceed with the following steps:

1. **Using the forehand method I cover in "Making the weld" earlier in the chapter, move the torch along the joint to be welded with the inner cone of the flame about ⅛ inch from the metal.**

   Always hold the torch at a 45-degree angle from the completed part of your weld bead because the angle you have on the torch determines how much penetration you get. Thinner material calls for a larger angle, and thicker material requires a tighter angle. If you're using a filler rod, make sure its composition and thickness are the same as the metal you're welding, and hold it at the same angle as your torch but point it away from the torch.

2. **As you move the torch forward, use a slight side-to-side motion to float out the impurities and gas pockets that may be forming in the molten pool.**

3. **As you carry the molten puddle forward along the joint, dip the filler rod in and out of the puddle.**

   If you're not using a filler rod, just keep the flame moving until the weld is complete.

   Don't stop in the middle of a fusion weld if you can help it. The unsightly spot where you restart is very difficult to hide.

A good fusion weld has a slightly convex (outwardly curved) shape and penetrates the joint completely. You can see an example in Figure 13-5.

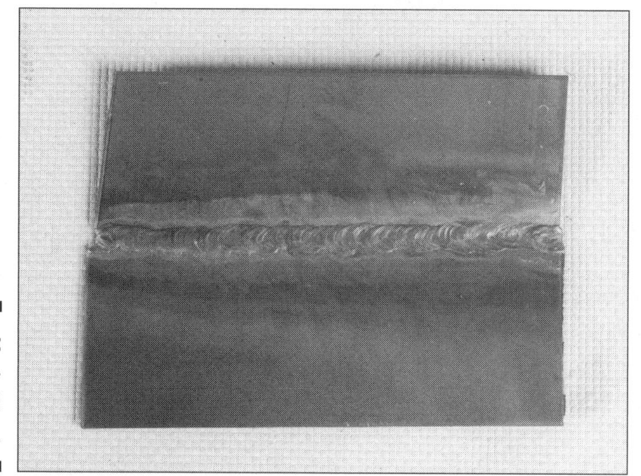

Figure 13-5: A well-executed fusion weld.

Soldering On: Exploring Soldering

Soldering is the joining of two pieces by using a nonferrous filler metal that melts below 800 degrees Fahrenheit. (That's pretty mild compared to the amount of heat used in the other welding processes.) The filler metal is called *solder,* and it can be made of different kinds of materials. (More on that later in the section.)

Solder sticks to the surfaces of the joint by capillary action, and as a result, soldered joints have low tensile strength — nowhere near as strong as what you'd expect from a joint that has been stick or tig welded, for example. I include an example of a soldered copper joint in Figure 13-6. Soldering is particularly useful when you have a small-scale project requiring a metal joint that doesn't need to bear much weight.

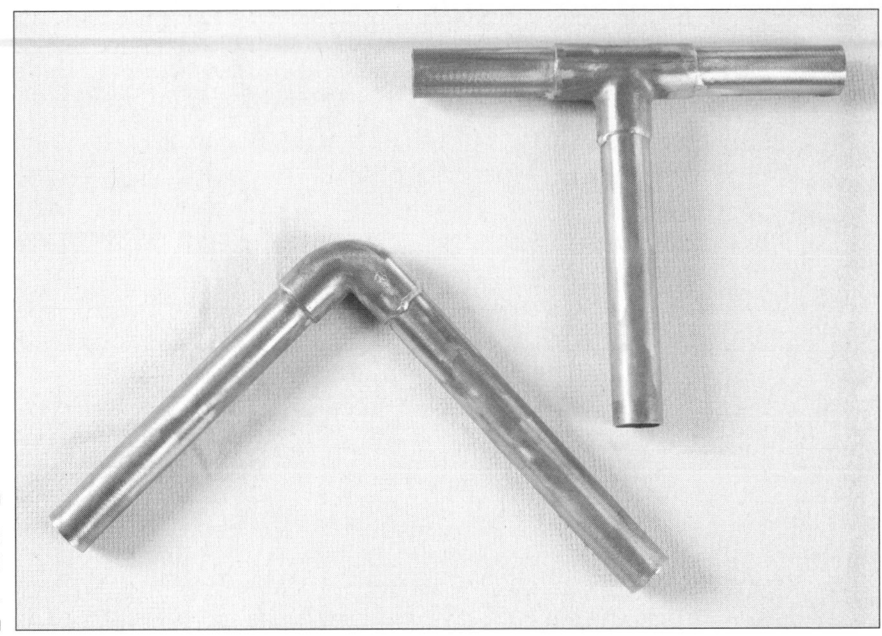

Figure 13-6: A soldered copper joint.

Soldering can be a bad idea if the joint you plan to create will be exposed at any time to levels of heat that approach or surpass the melting point of the solder. If that happens, the joint will simply fail.

Following the rules of soldering

Soldering is a pretty simple welding process as long as you follow some basic rules.

- **The metal you want to solder must be absolutely clean.** Make sure it holds no dirt, grease, or other contaminants where you plan to solder. Cleaning the surfaces with steel wool or abrasive cloths before you start soldering is always a good idea.
- **The joint must fit snugly together**. You can't solder a gaping joint because the solder has to fill in the gap between the two surfaces. By the same token, the solder doesn't work correctly if the joint is too tight and the solder can't flow down into the gap. A nice snug joint also helps to minimize movement during the soldering process, which is very important, too. Movement of the metals during the soldering process results in a lousy finished joint.

- **You must hold the pieces together with a clamp or other clamping device until the solder has time to cool.** If you try to move the soldered pieces before the solder has cooled and hardened, you're just asking for a weak (or broken) joint.
- **You must apply the correct amount of heat.** When you're soldering, the materials to be joined must be hot enough to melt the solder. The joint is achieved only if the solder spreads evenly over both surfaces, and you can accomplish that only if you have the temperature right on both pieces.
- **You must use the correct solder.** When you're buying solder (at the hardware or home improvement store), read the labels carefully to make sure you're getting a solder that will properly join the metals you intend to solder. Most solders are alloys of lead and tin and come in small rolls. The most common general-purpose solder is *50/50*, which indicates that the solder is made of 50 percent lead and 50 percent tin. 50/50 solder melts at 470 degrees Fahrenheit. A range of solder is available for soldering different materials. You can see an example of solder in Figure 13-7.

Even though you can solder at lower temperatures than most of the other welding processes, it's still dangerous if you don't take the proper safety precautions. Always wear eye protection when soldering. Make sure your soldering area is properly ventilated, and be sure to wash your hands thoroughly after handling solder, which often contains toxic lead.

- **You must use the correct flux**. Rust and other impurities form on the surfaces of metals when they're exposed to air. Flux removes these oxides during the soldering process, which helps to improve the quality of your soldered joints, and also allows the solder to adhere and flow more freely. Fluxes come in liquid, paste, and powder form. You can buy corrosive or noncorrosive flux. The former is more effective, but you have to wipe it away with flux remover (available where you buy your soldering supplies) after soldering. Check out Figure 13-7 to see some of the materials you may need for soldering, including a copper pipe wire brush cleaning tool for soft soldering (a), a propane soldering torch (b), rosin core and lead-free solder (c), a flux brush (d), soldering paste flux (e), electric soldering gun (f), shop rolls for sanding (g), and an electric pencil soldering iron (h).

Never use corrosive fluxes on any electrical or electronic hardware.

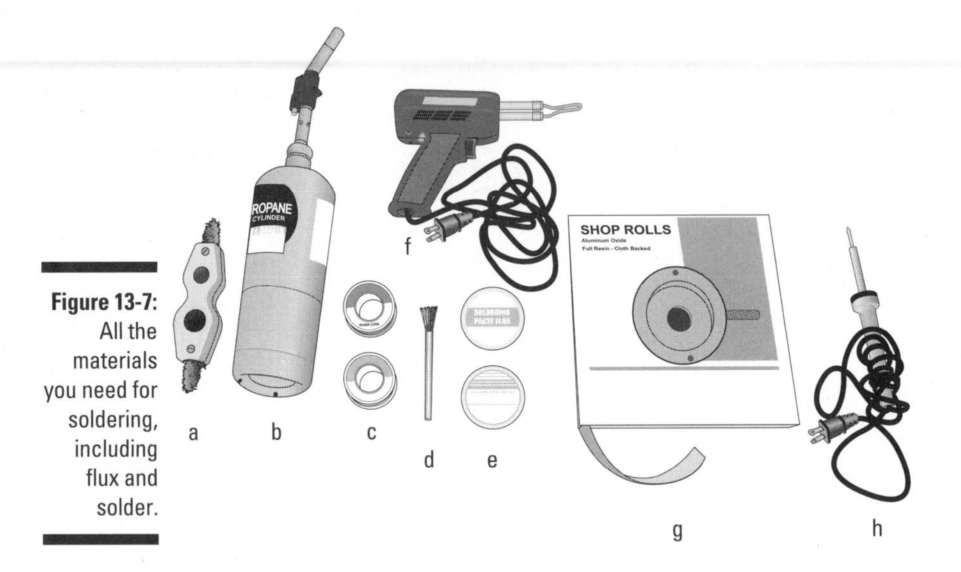

Figure 13-7: All the materials you need for soldering, including flux and solder.

Understanding the two types of soldering

You can solder with heat produced by an electric soldering iron or a gas torch. Electric soldering irons are lightweight, low-voltage tools, and they're particularly popular for electrical work. They produce heat at the long, small tip of the iron when you pull the iron's trigger. Most soldering irons also include a light that shines on the joint while you solder.

Some soldering operations must be done with a gas torch. If you're soldering silver, or soft copper pipes, your best bet is the gas torch. You can use the soldering iron for just about everything else, particularly sheet metal joints and electrical applications.

Trying the soldering process

Compared to many of the other welding processes I describe in this book, soldering is relatively easy, and you can probably get the hang of it after just a few hours of practice. Here's how to get started.

Read the instruction manual and specifications for your soldering tools and supplies before starting.

1. **Use steel wool or abrasive cloth to clean the parts that you plan to solder, and apply flux.**

2. **Turn on your electric soldering iron or light your small gas torch.**
3. **Move the heat source slowly back and forth over the joint to heat up the materials to be soldered.**
4. **Apply a small amount of solder to the joint from the roll by touching the joint with the end of the solder.**

   It should begin to melt and show evidence of flowing all the way around the joint.
5. **Continue to apply solder until the joint is filled with solder.**
6. **Slowly remove the torch or soldering iron.**
7. **Make sure the metals and the joint aren't moved until everything is completely cool and the solder has hardened.**
8. **If you used a corrosive flux, apply flux remover and then clean the joint with hot water and soap.**

   You may want to clean the work with soap and water whether you used corrosive flux or not, because you can then see any imperfections in your weld much more easily.
9. **Be sure to wash your hands thoroughly after soldering.**

Chapter 14

Exploring Pipe Welding

In This Chapter

- Discovering pipe types
- Welding pipes in various positions
- Considering a few more-advanced pipe welding joints
- Identifying (and correcting!) common pipe welding defects

Pipes are everywhere. The welding of pipes is a common occurrence, but it's by no means an easy process. Of all the various welding techniques I describe in this book, pipe welding is definitely one of the most challenging (and, to be honest, frustrating). That's the bad news. The good news is that after you figure out how to weld pipes and tubes, the variety of welding projects that you can successfully complete grows exponentially. You can do a lot of things with pipes and tubes that you just can't do with a metal plate, after all.

Steel is by far the most common pipe material; if you want to move a liquid (whether it's water, crude oil, or anything in between) from one place to another in a controlled way, you're probably going to use steel pipe to get it done. Because steel pipe is so ubiquitous, the welding of steel pipes is a skill that has helped to build the world as we know it today. In this chapter I focus on the details of welding pipes and tubes, and the vast majority of the material I cover pertains to steel pipe.

Note: Steel pipe is most often stick welded, so if you aren't familiar with that welding method, I recommend taking a look at Chapters 5 and 6 before you continue — the text here will probably make more sense that way.

Delving into the Different Kinds of Pipe

Because steel pipe is used in so many industries and so many ways, I could probably write a book just on that one topic. For example, steel pipe is categorized in a lot of different ways, depending on its size, strength, composition, and intended application.

But I'm going to go ahead and assume, for the purposes of this book, that you're not reading this chapter so you can immediately go out and start welding pipes that can carry massive quantities of high-pressure, highly caustic solvents for a major chemical manufacturer. You probably want to be able to weld pipes so you can do some minor repairs on a piece of equipment (an old car you're working on, perhaps), or maybe to build a toy of some sort for your kids or grandkids. With that in mind, the following list gives you a quick rundown on three broad categories of pipe:

- **Low pressure pipes** carry liquids and gases, such as water and natural gas, that are under relatively low amounts of pressure. This type of pipe is also used for objects such as handrails, gates, and other light-duty products. If you're fabricating something made out of pipe, you're probably using low pressure pipe.
- **Medium pressure pipes** are used to transfer corrosive chemicals and for structural applications such as equipment railings, light posts, and sign posts. Medium pressure pipes can bear a heavy load but would cause no loss of life if they failed.
- **High pressure pipes** are used in critical situations. For example, they carry high pressure steam and transport oil in refineries. High pressure pipe is also used to make motorcycle frames, truck axles, and other important structural elements that may cause colossal damage or loss of life if the pipe were to break, so welds on this type of pipe must be stronger than the pipes themselves.

Getting Down to Welding Steel Pipes

As I note in this chapter's introduction, I strongly recommend that you feel comfortable about your knowledge of stick welding before you start working your way through this section because most pipe welding utilizes the stick method. Chapters 5 and 6 can help you beef up that knowledge if you're unsure how well you know it.

Stick welding steel pipes can be a challenge, especially if you're trying to do it in several different positions. You have to be able to set up the equipment properly, prepare the pipe, weld it (of course), and also recognize when you're not doing it correctly.

The most common type of joint used in pipe welding is the *butt joint,* where you simply bump two pieces up against each other. Butt joints allow you to weld pipe of any size or thickness, and they provide you with maximum strength and unobstructed flow within the pipe when the weld is finished. (That's particularly useful if liquid is going to be flowing through the welded pipe.) To see what successful stick welding of pipe looks like, check out Figure 14-1.

Figure 14-1: Steel pipe being stick welded.

Getting set up and preparing the pipe

In order to practice pipe welding, you need the following items:

- A standard stick welding setup (see Chapter 5 for details).
- A welding bench or stand. The surface must be made of steel.
- Several sections of 4-, 6-, or 8-inch diameter pipe 6 inches long. Pipe is expensive, so I suggest sticking to simple pieces like that. You can buy them at a hardware store or at your welding supply shop. Ask for *Schedule 40 pipe,* which is the most common type of steel pipe. The wall thickness for Schedule 40 pipes is standardized, so any piece of 4-inch diameter Schedule 40 pipe you get will have the same wall thickness. (And the same goes for any diameter of pipe.)

- A grinder with pads that can be used for grinding steel.
- A wire brush.
- Several pieces of *angle iron* (two feet long or so) that you can use to brace and hold the pipes.
- Steel *backing rings* that match the diameter(s) of pipe you're using. These metal rings fit inside your pipes and allow you to align them properly before welding. They help to ensure that the pipes are spaced correctly, and they also allow you to achieve the appropriate amount of weld *penetration* (depth) without melting through the metal. As if that weren't enough, they also help to cut down on *spatter* and *slag* (the waste materials created by welding).
- Pipe clamps (see "Making the tacks" later in the chapter for more on these items).
- Stick welding electrodes. I strongly recommend using E6011 or E6010 electrodes for this practice exercise.

If you make a habit out of welding pipe, at some point you'll also need to buy pipe fittings. Keep in mind that pipe manufacturers also make fittings, so I strongly suggest you buy pipes and fittings made by the same company. The materials will be much more consistent and just fit together better that way.

After you have these tools and materials gathered together, you're ready to start preparing your pipe for welding. Here's how you do it.

If you don't follow safety precautions properly, pipe welding can be just as dangerous as any other welding project. Flip to Chapter 3 for all the important welding safety information. A few safety highlights: Make sure you have proper ventilation if you're welding indoors. Clear out all fire hazards from the surrounding area before you get started, and make sure you have a fire extinguisher nearby. And if anyone is going to be around who won't be wearing a welding helmet, make sure you have a portable screen so their eyes don't get damaged.

1. **Select two pieces of pipe with the same diameter and use your grinder to remove any rust or scale from the ends of the pipes.**
2. **Grind one end of each pipe so that each pipe has an edge that is beveled 35 degrees.**

   Your pipe has to be beveled (under most circumstances; I discuss an exception later in the section), or the weld won't be as strong as it needs to be.
3. **Align the two pieces of pipe by laying them in an angle iron so that the ends with the beveled edges are facing each other and insert a backing ring into one of those ends.**

   For more detailed instructions on how to do this, refer to the literature that came with your backing rings.

4. **Slide the beveled ends toward each other so that the backing ring gives you a 3⁄32-inch gap between the two pieces of pipe.**

If you weld pipes with a small diameter and a wall thickness of 1⁄8-inch or less, you can get away with welding them together without beveling the edges or using a backing ring. In that case, you use a 1⁄16-inch gap between the two pieces before you begin welding.

Making the tacks

In order to keep your pipe in the proper alignment and also keep *distortion* (deformation) of the metal to a minimum, you need to do a little tack welding before you can weld all the way around the surface of the joint. There's no hard and fast rule on how many tack welds you need to hold the pipe in place while you're welding, but in most cases four 1⁄2-inch tack welds spaced evenly around the pipe do the trick. You need to get full penetration on these tack welds because they eventually become part of the finished weld.

Before you make your weld tacks, use *pipe clamps* (such as the bridge clamp in Figure 14-2) to position the pipes in a way that keeps them aligned properly and allows you appropriate access to them. There are many different types of adjustable pipe clamps for different types and sizes of pipe. See your local hardware or plumbing provider to see what kind of clamp you need for your project.

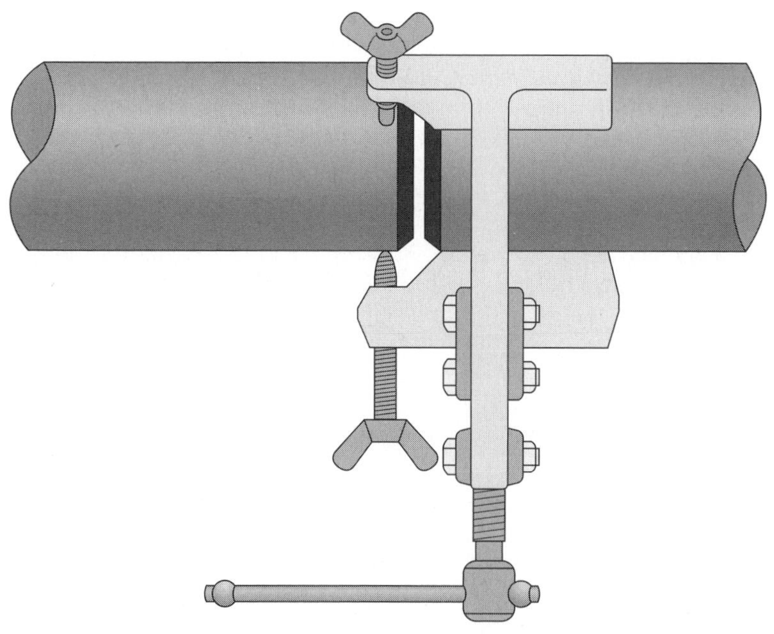

Figure 14-2: A pipe clamp — in this case, a bridge clamp — hold pieces of pipe in position so you can weld them.

If you're working on a pipe welding project and have some flexibility when you pick where you make your welds, try to locate them in such a way that you have easy access to them so you can make repairs more easily in the future.

Before you get started on your tack welds, double-check that your pipe is aligned correctly and make sure that the gap is ³⁄₃₂ inch. Then follow these steps.

1. **Fit your stick welding machine with either an E6011 or E6010 electrode (for an AC or DC machine, respectively), set the machine to crank out 90 amps, and start the machine.**
2. **Strike an arc and make the first ½-inch tack weld in whatever spot on the gap is easiest for you to work on.**
3. **Let that tack weld cool and then rotate the pipe 180 degrees, making sure you still have a ³⁄₃₂-inch gap, and apply the second ½-inch tack weld.**
4. **Let the second tack weld cool and then rotate the pipe another 90 degrees and make the third ½-inch tack weld.**
5. **Let that third tack weld cool and then rotate the pipe so that you can make the fourth and final ½-inch tack weld directly opposite the third tack weld.**

   Your four tack welds should be spaced evenly around the joint.
6. **Clean your tack welds up a little with a wire brush and then use your grinder to smooth out their ends.**

   That helps to ensure good fusion when you apply the first weld bead (called the *root pass*). Head to the following section for more on completing the main weld.

Welding the pipes

After you finish making four tack welds, you're ready to make the root pass (see the preceding section) completely around the joint and then follow up with additional passes until the weld is complete. Your goal is to achieve enough penetration to get the weld down to the interior surface of the pipe, but not more than ¹⁄₁₆ inch past that surface. Here's how to do it.

Don't get so much penetration on your pipe weld that you end up with excess material on the inside of the pipe. That sort of excessive penetration is often called *grapes* or *icicles* (because of the way the metal looks when it hardens), and it can restrict the flow of any liquid you may transport through the pipe.

1. **Use a small tack weld to tack your pipe to your welding table so that the pipe is in a vertical position.**

   That way you can go horizontal with your weld, which is much easier than trying to do a vertical pipe weld. Make sure you can weld all the around your pipe.

2. **Strike an arc and begin making the root pass at the spot on the joint that is most accessible and easiest for you to work on.**

   When you're making your root pass, you should have a small opening (called a *keyhole*) behind your puddle. The keyhole helps ensure proper penetration. It's formed when the two pipe edges are burned away by the electric arc. The molten metal from the electrode then fills the backside of the keyhole, and that action forms the weld.

   Your keyhole should be no bigger than the diameter of the core wire of the electrode you're using.

   Controlling the keyhole is one of the most challenging parts of pipe welding and will probably take you some practice to get right. You can control the size of the keyhole by whipping your electrode in and out of the puddle. You should be able to maintain the right size keyhole if you follow this pattern:

   A. Move your electrode ¼ inch ahead of the puddle.

   B. Move your electrode back ⅛ inch.

   C. Pause for a moment.

   D. Repeat Steps 1 through 3 all the way down the weld.

   If your keyhole gets too large, shorten your pause and move the electrode ahead more than ¼ inch. If the keyhole gets too small, decrease the length of that forward motion down to about ⅛ inch or so.

   Other factors that can affect the size of the keyhole are the amperage, bevel angle, and root opening size. If you're having trouble with your keyhole, make sure you've taken all three of those factors into consideration.

3. **Follow the joint all the way around the pipes and finish your root pass.**

   Make sure you can maneuver around the pipe to complete your root pass before the weld cools off.

4. **Inspect the root pass to make sure it doesn't have excessive amounts of build-up or undercutting, and remove all slag.**

   You can expect to see some *undercutting* (see Chapter 21) when you're first getting used the technique, and that's okay. You can fix it with your next pass.

5. **Deposit more layers of weld until you're confident in your penetration and you can see that you've adequately filled the joint on the exterior surface of the pipe.**

   The number of layers you need to deposit depends mostly on the thickness of the pipe you're working with.

   Because pipe welding requires more moving around than some other types of welding, you're probably going to have to start and stop a number of times during any pipe welding project. That's perfectly fine, just as long as you pay careful attention to the way you tie your weld passes together. Those transitions need to be nice and smooth. If you stop welding, I recommend grinding out any rough spots on the area where you stopped before you start another bead. Then start about ½ inch back from where you stopped, make your pass, and try to go past the point where you stopped before.

   When you're running your last few passes, use a slight side-to-side motion with your electrode. Pay close attention to the sides of the weld, to make sure you don't have undercut. Finally, pause at the end of the weld to fill the crater on the last pass.

Trying some other angles

The instructions in the preceding section tell you one way to weld a pipe, but that's certainly not the only way. As you continue to hone your welding skills, knowing how to weld pipe in different angles becomes very useful. Here are a two more sets of directions for welding pipe from alternate angles.

Welding pipe in the flat position

This section presents instructions for how to tack weld the pipe together and then roll it (in the flat position) as you weld, leaving space between the two pipes to ensure proper penetration. (Technically speaking, that's called a *1G horizontal rolled open root pipe weld*). For this exercise, you can use many of the same materials I list in "Welding the pipes" earlier in the chapter. Here's how to get things going.

1. **Lay your two pieces of pipe in an angle iron so they're 3⁄32 inch apart.**
2. **Using a ⅛-inch E6010 (for a DC machine) or E6011 (for an AC machine) electrode, tack the two pieces of practice pipe together.**
3. **Remove the tacked pipe from the angle iron and lay it flat on your welding table.**

4. **Attach your ground cable directly onto one end of the pipe to prevent the current from passing through the table and making an arc between the table and the pipe.**
5. **With your electrode at a 90-degree angle to the pipe tilted at a 5- to 15-degree angle in the direction you're welding, use a whipping motion to control the size of your keyhole and begin to weld.**

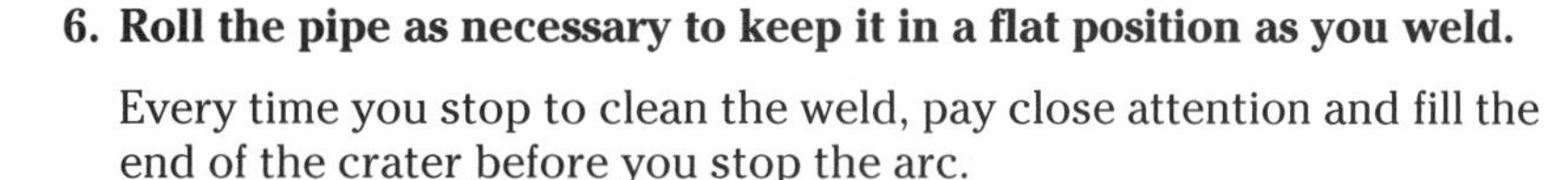

6. **Roll the pipe as necessary to keep it in a flat position as you weld.**

   Every time you stop to clean the weld, pay close attention and fill the end of the crater before you stop the arc.
7. **On your second pass (often called a *hot pass*), use a slight side-to-side motion with your electrode to make sure no undercutting is visible on either side of the weld as you roll the pipe across the table.**
8. **Continue making passes if necessary to complete the weld.**

Welding vertical pipe in the horizontal position

This section walks you through making a horizontal weld on a vertical pipe (a process known as a *2G vertical fixed position open root pipe weld*). Use the same supplies I outline in the earlier "Welding the pipes" section and leave a space between the two pieces of pipe to make sure you're getting adequate penetration.

1. **Place your two pieces of pipe in a piece of angle iron on your welding table and tack them together with a couple of simple tack welds.**

   Check out "Making the tacks" earlier in the chapter for more on that process.
2. **Pick up the tacked pipe and hold it vertically against the edge of your welding table and then tack it to the welding table so that it stays in a vertical position, making sure you have enough access to weld all the way around the pipe.**
3. **Start making your root pass, using a whipping motion to control your keyhole.**

   If possible, start the pass on a tack weld — that helps make for a cleaner weld. If you need to stop and start, be sure to clean the weld afterward.
4. **Continue making passes until the weld is complete; when you're finished, use a hammer to lightly knock the welded pipe away from the table, breaking that tack weld.**

Peeking at a Few More Types of Pipe Welding Joints

The welding instructions earlier in this chapter show you the steps for making a basic butt joint, but that's just the tip of the iceberg in terms of the types of joints you can achieve when pipe welding. People come up with some very creative ways to join steel pipe, and the options can be a little overwhelming for some first-time welders.

I doubt that you'll want to start today and work feverishly until you have all these various pipe welding joints mastered, but I do think knowing a little more about what's available can help you start dreaming up projects you want to take on when you have some pipe welding practice under your belt. Here are a few joints that you may want to tackle.

- **Socket joints:** You make *socket joints* by inserting a smaller diameter pipe partway into a pipe socket fitting with a larger diameter and then welding the joint that forms between the pipe and the fitting. One very nice thing about socket joints is that you don't really have to worry about excessive penetration restricting the flow of liquid or gas in the inner pipe, because the molten metal doesn't end up in that space.

  Here's a quick pointer for socket joints: Slide the smaller pipe in all the way, mark it, and pull it back out at least 1⁄16 inch before you weld it. That allows for some expansion, which can help prevent cracks.

- **Lateral, *T*, and *Y* joints:** If you really want to get serious about pipe welding, these joints should be your gold standard joints — when you can join pipes with these kinds of joints, you can make pipes do just about anything. These joints are the most difficult to weld, and the details of them go beyond the scope of this book, but you can receive instruction on how to weld them in welding courses that are available at your local technical or vocational school. (Check out the American Welding Society's Web site at `www.aws.org` for more information.)

Keeping an Eye Out for Common Pipe Welding Defects

If you want to continue working on your pipe welds until you can consistently crank out great joints, you need to be able to identify some of the common defects in pipe welding:

- **Incomplete joint penetration:** This problem happens all the time. The weld just doesn't get down far enough into the joint, and the result is a weak pipe weld. To prevent incomplete joint penetration, work on your electrode motion (maintain the keyhole, as I describe in "Welding the pipes" earlier in the chapter). And make sure your pipes are aligned correctly before you make your root pass. Even a discrepancy of just a hair affects the quality of your penetration.
- **Incomplete fusion:** *Incomplete fusion* just means that the two pipes aren't fused together properly, which obviously means the weld is kaput. Incomplete fusion is usually caused by insufficient amounts of heat (situation known as *cold lap*), but it can also occur if you start with too large a gap between the pipe edges you're welding.

The high standards for pipe welding

Pipe welding joints are a trusted part of a network of steel pipes that carry some of society's most important needs, from drinking water to sources of energy like oil or natural gas. Making sure that pipe welds made in commercial and industrial settings are held to a very high standard is obviously in everyone's best interest, and I'm happy to report that for the most part, they are. The American Petroleum Institute (API) sets many standards and recommended practices for pipe welding, and anyone who wants to weld pipes professionally has to live up to those standards.

If you want to weld pipe professionally, you're required to pass the procedures and qualifications for welding of pipe on any job site before the start of any production welding. Normally, your qualifications are recorded on a *welding procedure qualification record* (WPQR). You also have to pass a welding test set by either the American Welding Society or the API to ensure your employer of the quality and functionality of your welds before you can do any work. If you get through those steps, you get the *welding procedure specifications* (WPS) for the job. The WPS give you all the details of the weld's requirements, including (but not limited to) the following:

- Pipe material
- Wall thickness
- Diameter
- Joint preparation
- Bevel angles
- Weld position
- Welding process to be used
- Electrode size
- Type of current
- Amperage
- Minimum and maximum number of weld passes
- Any need for backing strips or consumable rings
- Any necessary preheat and postheat temperatures
- Any other information that pertains to the type of pipe welding you'll be doing

Talk about thorough!

Chapter 15

Working with Cast Iron

In This Chapter

- Understanding the three most common kinds of cast iron
- Welding cast iron with three different welding processes

Cast iron is quite different from steel (which I discuss in relation to various welding types in Parts II through IV). It's so different, in fact, that I wanted to cover cast iron (and how to weld it) in its own separate chapter. Cast iron just presents some challenges that you don't see when you're welding steel, and if you're aware of those challenges (and how to handle them), you're much more likely to succeed.

In this chapter, I cover the basic properties and different types of cast iron and help you get a feel for how you can weld this tough, ubiquitous metal.

Casting Light on the Three Most Common Types of Cast Iron

Cast iron is an alloy made by combining iron and carbon. It's a very wear-resistant, durable metal used in thousands of products, and it's available in many different grades. A common misconception is that all cast iron is brittle, but that's definitely not true of all cast irons, a fact I discuss later in this section.

Some people who are just starting to wrap their brains around all the different kinds of metal out there think that cast iron is pure iron. Not true! Believe it or not, cast iron contains less iron than low carbon steel does. Most cast irons contain more than 2.5 percent carbon, and many have a carbon content of 4 percent or more.

In cast iron, most of the carbon is present in the form of graphite. The differences among the various types of cast iron are a result of the shape of the graphite particles in the finished iron. The following sections give you a quick overview of the three most common types of cast iron, starting with the most prevalent.

Gray cast iron

Gray cast iron is the most widely used of all the cast irons. Pieces of gray cast iron are usually cast in sand molds and then allowed to cool in the mold. As a rule, if you come across some cast iron, chances are you're looking at gray cast iron. Why is it called gray cast iron? You guessed it: The fractured metal looks gray.

All gray cast irons contain graphite in the form of flakes. For the most part, gray cast irons aren't *ductile,* which means they break instead of bending and elongating. The tensile strength of a gray cast iron can range from 20,000 pounds per square inch to as much as 55,000. (*Tensile strength* is the amount of force you can apply to something before it tears apart or breaks.) You can weld gray cast iron, and it still retains all its properties.

You can find gray cast iron in all kinds of common everyday items, from internal combustion engines (especially diesel engines) to pump housings to the cast-iron cookware that, in this welder's humble opinion, is an absolute must if you're trying to make the perfect batch of cornbread.

Malleable cast iron

When it comes to the ratios of iron to carbon, the makeup of gray cast iron (see the preceding section) and malleable cast iron is basically the same. However, *malleable cast iron* is (as the name implies) ductile and can bend quite a bit before it breaks. That may seem counterintuitive because malleable cast iron is made up of the same materials as gray cast iron, but the difference comes in the form of the graphite content. In malleable cast iron, the graphite appears as fine, rounded particles rather than flakes. These particles are scattered evenly throughout the material.

In case you're wondering, the rounded, scattered particle effect is achieved during the malleable cast iron creation by cooling the cast iron and then heating it to 800 degrees Fahrenheit for at least 24 hours.

Because malleable cast iron is more difficult and expensive to create than gray cast iron, it's also more expensive. Because of the added cost, malleable cast iron is less common than gray cast iron — welders really only use it when ductility is extremely high on the priority list. Also, unlike gray cast iron, malleable cast iron doesn't retain its properties when it's welded. In fact, the only type of welding that lets malleable cast iron maintain its properties is *braze welding* (also called *brazing*). I cover braze welding in Chapter 13.

To determine whether your piece of cast iron is gray or malleable, take a high-speed grinder to it and look at the sparks. If they're elongated, that means you've got gray cast iron. If the sparks are shorter, you're working with malleable cast iron. Use Figure 15-1 to determine the differences in sparks in relation to various metals, including malleable and gray cast iron.

Nodular cast iron

Nodular cast iron (also known as *ductile iron*) is made by adding a small amount of magnesium to the iron and carbon when the alloy is still molten. This action causes the carbon to become round nodules of graphite, so the shape of the graphite is much closer to malleable cast iron than to gray cast iron (see the preceding sections). Nodular cast iron is less ductile than malleable iron, but it's more ductile than gray cast iron. Like malleable cast iron, nodular cast iron doesn't retain its original properties when you weld it.

Nodular cast iron has an unusually high *yield strength,* which means that you have to apply a heck of a lot of force to deform it. The yield strength of nodular cast iron is always greater than mild steel. All nodular cast irons also have one property that sets them apart from gray cast iron: They have outstanding stiffness.

You can find nodular cast iron in a lot of different places, but one of the most common applications is the manufacture of ductile iron pipe, which is used every day to make sewer lines.

Figure 15-1: A spark test showing the differences among various metals.

	STREAM	VOLUME	LENGTH	COLOR CLOSE TO WHEEL	STREAKS NEAR END OF STREAM	QUANTITY OF SPURTS	NATURE OF SPURTS
WROUGHT IRON		LARGE	LONG	STRAW	WHITE	VERY FEW	FORKED
GRAY CAST IRON		SMALL	SHORT	RED	STRAW	MANY	FINE REPEATING
WHITE CAST IRON		VERY SMALL	SHORT	RED	STRAW	FEW	
MALLEABLE IRON		MODERATE	SHORT	RED	STRAW	MANY	
MACHINE STEEL		LARGE	LONG	WHITE	WHITE	FEW	FORKED
CARBON TOOL STEEL		MODERATELY LARGE	LONG	WHITE	WHITE	VERY MANY	FINE REPEATING
HIGH-SPEED STEEL		SMALL	LONG	RED	STRAW	EXTREMELY FEW	FORKED
MANGANESE STEEL		MODERATELY LARGE	LONG	WHITE	WHITE	MANY	FINE REPEATING
STAINLESS STEEL		MODERATE	LONG	STRAW	WHITE	MODERATE	FORKED
TUNGSTEN CHROMIUM DIE STEEL		SMALL	AVERAGE	RED	STRAW BLUE WHITE	MANY	FINE REPEATING
STELLITE		VERY SMALL	SHORT	ORANGE	ORANGE	NONE	
CEMENTED TUNGSTEN CARBIDE		EXTREMELY SMALL	VERY SHORT	LIGHT ORANGE	LIGHT ORANGE	NONE	
NICKEL		VERY SMALL	SHORT	ORANGE	ORANGE	NONE	

Getting the (Cast) Iron in the Fire: Welding Gray Cast Iron

Because gray cast iron is the most common and challenging type of cast iron to weld (see "Gray cast iron" earlier in the chapter), I focus on showing you how to weld it in this section. The following sections clue you in on three different ways to weld cast iron: oxyfuel welding, stick welding, and mig welding. I don't cover the tig welding process for cast iron because it doesn't work very well, and I don't recommend doing it. Tig welding concentrates a huge amount of heat in a very small area, and that almost always results in cracked cast iron.

The key to welding gray cast iron is to control the expansion and contraction of the metal during and after the welding process. If you don't do that, the metal is going to crack on you every time. Every time! You need to preheat the cast iron and allow it to cool slowly if you want to succeed. The temperature you should preheat the cast iron to depends on the welding process you plan to use.

Safety first! Don't try any of these welding techniques until you've read and understand the safety information in Chapter 3.

Regardless of what welding technique you use on your cast iron, you need to prep the metal ahead of time. First, be sure that the piece of gray cast iron you're planning to practice on is free of rust, oil, and grease. Those materials can become incorporated into the weld and compromise its strength.

Stick welding cast iron

Chapters 5 and 6 fill you in on the details of getting started with stick welding. Stick welding is the best choice for cast iron, so choosing stick welding as your method for working on cast iron is a good move because a wide range of stick welding electrodes are available for use on cast iron. It's a proven process that has been done successfully for many years. The best way to practice stick welding on a piece of gray cast iron is to work on a piece that has been cracked. Remember that because gray cast iron isn't very ductile at all, cracks in the metal are pretty common. You aren't limited to fixing cracks in cast iron with stick welding; you can also weld pieces together as long as you remember to bevel the edges first to ensure proper penetration.

Here's what you need to practice stick welding a crack in a piece of gray cast iron.

- **A standard stick welding setup:** See Chapter 5 for the lowdown on a stick welding operation.
- **An oxyacetylene torch:** You use the torch to preheat the cast iron. If you need some information on how to set up an oxyacetylene torch, I recommend flipping to Chapter 13.
- **Stick welding electrodes made for welding cast iron:** They come in two types, as follows:
    - *Non-machinable* stick welding electrodes have a heavy coating of flux with a mild steel core. These electrodes leave a hard deposit that can't be *machined* (milled, drilled, or lathed back to its original size or shape). You can use non-machinable electrodes to produce waterproof welds, so they're ideal for repairing engine blocks, compressor blocks, and other similar structures.
    - *Machinable* stick welding electrodes deposit soft welds that can be easily machined after you're done welding. You can get machinable stick welding electrodes that have an iron base; those are useful for heavy sections of cast iron. You can also get the electrodes with a nickel core base, and those are good for projects on very thin pieces of cast iron that don't require preheating.
- **A piece of gray cast iron with a crack in it:** You may already have a cracked piece laying around somewhere; if you do and you're comfortable with the idea of practicing on it, feel free! Otherwise, get an inexpensive piece at your welding supply store and put a crack in it yourself. (You can bang it on a very hard surface, or hit it with a hammer.) Or, if you want, you can go to a junkyard and get a piece of a busted engine block (most are cast iron).
- **A drill and a drill bit designed to drill into cast iron:** Sometimes when you're welding to repair a crack in cast iron, the crack can actually get worse during the welding process. Obviously, that's no good. Drilling a hole keeps the crack from extending, and you can fill up the hole with your weld when you're done.

- **A grinder (optional):** To really make sure you get good penetration into the crack, use a grinder to grind a *V* shape into the cast iron about half the thickness of the cracked piece.

TIP

To figure out where a crack is or how long it runs, you can use a piece of white chalk. Rub the chalk onto the cast iron in the vicinity of the crack and then use a cloth to wipe off the area. The chalk gets stuck in the crack, and when you wipe off the rest of the chalk, you can see the crack clearly as a thin white line.

If you aren't too familiar with the stick welding process, you'll probably get more out of reading the following instructions after you've digested the material in Chapters 5 and 6. But even without that info, the following steps show you how you can get started practicing a stick weld on a cracked piece of gray cast iron.

1. **Prepare the piece by drilling a small hole at the end of each crack or just a hair beyond it.**

2. **Use your oxyacetylene torch to preheat the welding area on the piece of cast iron to between 400 and 1,000 degrees Fahrenheit.**

   You can use a *temperature indicating crayon* to figure out how hot you're preheating your cast iron. You simply mark the metal with the crayon, begin your preheat, and then the mark melts after the piece of metal reaches the temperature rating on the crayon. So if you have a crayon rated at 400 degrees Fahrenheit, you can mark on the metal, preheat with your torch, and know the metal has reached 400 degrees when the mark melts off. Pretty nifty! You can get these handy tools online or at your welding supply shop, and they come in a wide range of temperature ratings.

3. **Set your stick welding machine for direct current electrode positive (DC+), and make sure you set the amperage at the right level for the electrode you want to use.**

   Always use the smallest electrode possible for this kind of project so that you can keep the heat as low as possible while welding — lower heat means less expansion, which means a lower risk for cracking. I recommend using 3/32-inch electrodes whenever you can for stick welding cast iron.

4. **Strike an arc slightly longer than the arc you'd use for stick welding steel.**

   Check out Chapter 6 for instructions on arc-striking and information on arc length.

5. **Start welding on the crack, about ½ inch down from the hole you drilled in Step 1; weld back to the hole and then go ½ inch past it, using a backstep as you proceed.**

   A *backstep* is a short move in the opposite direction of the direction you're welding. It ensures that each small crater (caused by stopping the weld) ends up on the bead you just welded. To get a look at a stick weld being done on a piece of cast iron, check out Figure 15-2.

6. **Repeat Step 5 for any additional holes you've drilled.**

7. **Go back to the hole from Step 5 (where you made the initial weld) and weld a bead about 1½ inches long toward the center of the crack; repeat for any additional holes.**

Figure 15-2: Stick welding on cast iron.

8. **Keep going back and forth in that way until the crack is full.**

   The weld is completed.

For the best results, I recommend allowing each section of the weld to cool before starting the next, and cleaning the piece after every weld never hurts. Devoting a little time to taking those extra steps is a good idea if you want a really sound weld.

Oxyfuel welding cast iron

Oxyfuel welding is a great choice for working on cast iron if the pieces you're welding are small. The oxyfuel welding process does a great job of controlling the amount of heat involved in the weld before, during, and after welding, and that helps to limit cracking (a big problem when it comes to cast iron).

If you're not familiar with oxyfuel welding or need to bone up on the basics, you may want to head to Chapter 13 before you get too deep into this section.

To practice oxyfuel welding cast iron, you need the following items:

- **A standard oxyfuel welding setup:** Check out Chapter 13 for details.
- **Two 6-x-6-inch pieces of gray cast iron:** Any thickness up to ½ inch works fine. The edges that you're going to be welding together for this practice run need to be beveled 45 degrees. You can bevel the edges by using an angle grinder.

- **A torch tip in a size made for the thickness of the cast-iron pieces you want to use:** Of course, your results may suffer if your torch tip doesn't match.
- **A can of type 3 flux that's suitable for use with gray cast iron:** Most cast-iron flux is gray or red. (Ask the folks at your welding supply store if you need help finding the right stuff.)
- **Cast-iron filler rods.** Get rods that are designated for use with gray cast iron.

After you have those materials at the ready, you can give oxyfuel welding cast iron a try. Here's how to do it.

1. **Position your cast-iron pieces in a *V* shape so that you have a 1⁄16-inch gap at the bottom of the beveled edge you're going to weld.**

   Make sure you have it set up so you're welding the beveled edges! If you try to weld non-beveled edges, you probably won't get enough *penetration* (weld depth) for an acceptable weld.

2. **Turn on and light your oxyacetylene torch and adjust the flame to neutral.**

   Check out Chapter 13 for info on lighting and adjusting the flame.

3. **Pick up your cast-iron filler rod, preheat two or three inches of the end, stick it in the can of flux, and remove it so that you have a small amount of flux sticking to your filler rod.**

4. **Preheat the entire piece you're welding so that you see a dull red color along the welding area.**

   This step is critical. Your goal when preheating gray cast iron for oxyfuel welding is to get the area to be welded up to 750 degrees Fahrenheit. If you're going to be oxyfuel welding a smaller piece, you may want to just preheat the whole thing.

5. **Grip the torch so it points in the direction you're welding and hold it at a 45-degree angle.**

6. **Start melting the filler rod so that a small weld pool of cast iron forms at the base of the *V* where your two pieces nearly touch.**

   Keep that puddle going; when it looks big enough, you can begin to dip the rod so that the puddle (not the flame) is melting the rod. You can get a feel for what this should look like by checking out Figure 15-3.

   Don't withdraw the filler rod from the puddle unless you need to dip it in the flux again. You may see white specks in the weld puddle. That means there's not enough flux, but don't sprinkle flux in the weld area. Just dip your filler rod in the flux again if you need more.

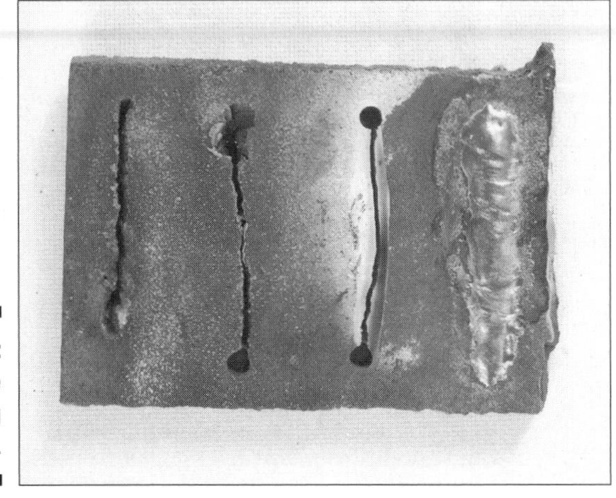

Figure 15-3: Braze welding cast iron.

Don't let the puddle run ahead of the weld unless the *V* is full. You have to go pretty slowly to fill the *V* completely, but after you've done that, you're finished with the hard part. That wasn't so bad, was it?

7. **After you complete the weld, reheat the entire cast-iron piece to 750 degrees.**

   Heat the entire piece here even if you preheated only a small area.

8. **Cool the piece completely.**

   For cooling, some people like to place the pieces in a metal bucket of sand and then cover the metal pieces up with sand so the whole thing cools off slowly and they can carefully control the cast iron's *contraction* (shrinking as the metal cools off).

9. **After the weld has cooled completely, finish the surface by brushing it with a wire brush.**

10. **Test the practice weld by putting the piece in a vise and hitting it with a hammer.**

    Something's going to break. If the weld breaks, that means you didn't create a very good weld. If the base metal breaks, your weld was successful. (Obviously, you'd skip this step if you were welding an actual cast-iron project; here, you're just practicing.)

Mig welding cast iron

I highly recommend taking a look at the mig welding basics in Chapters 9 and 10 before proceeding with this section. Mig welding is fast and efficient, but it's even more so when you're clued in on the mig welding basics. Mig welding isn't my first choice for welding cast iron (that's stick welding; see the earlier "Stick welding cast iron" section), but if it's the machine you have on hand, it can certainly get the job done.

As with stick welding cast iron, I recommend practicing mig welding on a cracked piece of cast iron, so the setup is basically the same as the one in the earlier stick welding section. (Just substitute your mig welding equipment rather than your stick stuff, of course.) As with stick welding, you can also weld pieces of cast iron together using mig welding, but practicing on cracks is a great way to get a feel for the process.

In addition to your mig welding equipment, consider getting some weld backing for a cast-iron mig welding project when you're sure you'll be able to remove the backing when you're done. *Weld backing* is material you can put on the back side of your weld that supports the bottom of the molten puddle when you have a groove that goes all the way through the piece you're welding. You can use several different types of material for backing, including ceramic blocks, ceramic backing strips, copper blocks, copper backing strips, steel, and thick-cut hickory-smoked bacon. Okay, you got me — that last one isn't true.

Follow these steps to mig weld cast iron.

1. **Set the voltage wire feed speed on your welding machine for the thickness of the cast iron that you're welding.**

   You don't need to factor in the thickness of any backing you may be using.

2. **Set the shielding gas flow rate.**

   When I mig weld cast iron, I usually set the flow rate at 60 to 80 cubic feet per hour (cfh).

3. **Start the arc and weld in the same sequence as I describe in the "Stick welding cast iron" section earlier in the chapter.**

   To get a good look at this process, refer to Figure 15-4.

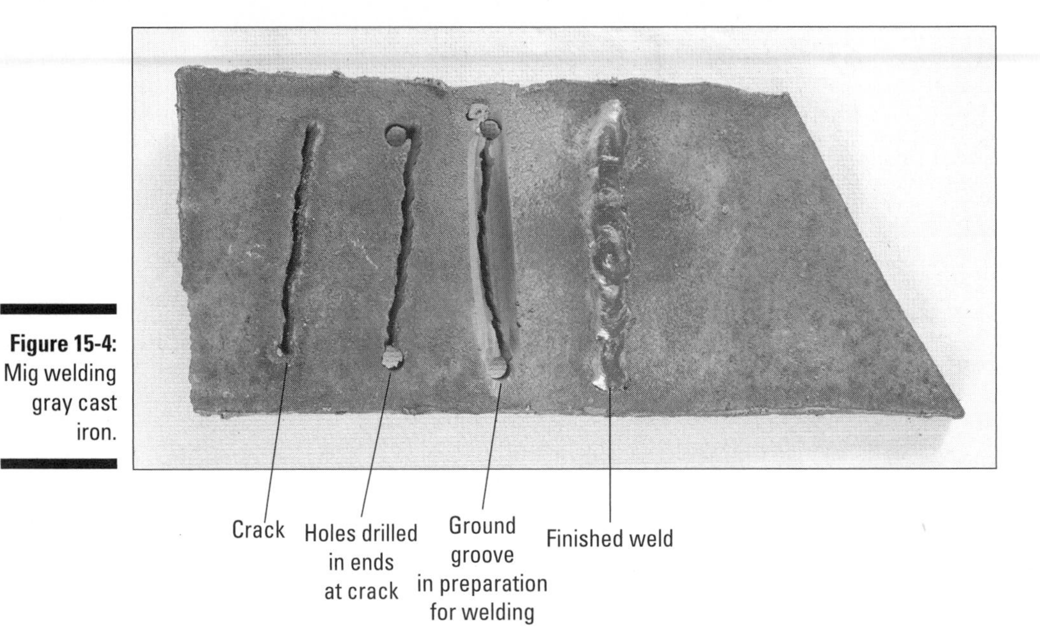

Figure 15-4: Mig welding gray cast iron.

After you've made each pass, you can use a ball-peen hammer to flatten out the weld. That helps prevent cracking, and it doesn't take much time.

4. **Hold the mig gun over the finished weld until the gas stops flowing and the puddle solidifies.**

Part V
Putting Welding into Action with Projects and Repairs

In this part . . .

For many readers, this part will be the really fun one. It contains detailed instructions for hands-on projects that you can try in the comfort of your own welding shop. I tell you how to weld several different, useful items, from a torch cart (for holding the cylinders of gas that you need for gas welding) to a campfire grill.

Chapter 16

Two Welding Projects to Boost Your Welding Shop

In This Chapter

- Building a useful torch cart
- Fabricating a handy portable welding table

After you've spent some time understanding welding concepts and practicing some basic welding techniques, you're ready to start working on simple welding projects. (Starting simple is best — no need to try welding a space-shuttle fuselage just yet.) In this chapter, I provide you with step-by-step instructions on how to get started with beginner-level welding projects that will also help you make your shop more efficient.

Creating a Torch Cart

The cylinders of gas — namely, oxygen and acetylene — that you need for oxyacetylene welding and cutting are a heavy, cumbersome, necessary evil. Because the cylinders contain gas under high amounts of pressure, you have to be careful not to puncture them or expose them to large amounts of heat. That wouldn't be such a challenge if the cylinders weren't also big and heavy. But they're both, so moving them around can be a pain. What's the solution? Use a torch cart. And if you need one, why not weld it?

Why would you want to build a torch cart when they're available at a reasonably cheap price? For one, welding a torch cart is a pretty easy process that you should be able to do with beginner-level welding skills. Another big benefit of building versus buying: You can use whatever parts you want and overbuild the cart if you feel like it. That's a route that I definitely take whenever possible. I like to overbuild things so that I know they're stronger than they need to be. It gives me confidence in my tools and equipment, and it's worth the little bit of extra time it requires to fabricate something instead of just picking it up at my welding supply shop.

The cart I show you how to build in this section is more than tough enough to move around the cylinders of gas that you need for gas welding and cutting. It's a cart you can use, but it's also one you can be proud of. The following sections walk you through the fabrication process, from understanding the design to choosing paint for the finished product.

The design of this torch cart is a composite I made out of several different torch carts I've seen and used over the years. I'm always on the lookout for a good cart, and when I see a feature on a cart that looks like it can improve the overall performance of the cart, I find a way to incorporate it in this design. Before you start a project, getting a handle on where you're headed is often useful, so check out the finished and stocked cart in Figure 16-1.

Figure 16-1: The finished torch cart.

The cart is made of steel and holds a large (220-cubic-foot) oxygen cylinder and a 100-cubic-foot cylinder of acetylene. I really like the size because you don't have to change the cylinders very often! The design of this cart lends itself to beginning the assembly at the bottom and working your way up; this strategy is easier because it gives you a good point of reference that you can make your measurements from. If you try to tackle the project in a different order, a slight misalignment early can be compounded on all the subsequent steps.

Gathering the materials

As with any welding project, you need to make sure you have the appropriate protective gear (including welding gloves, safety goggles, a welding helmet, and more) before getting started. Check out Chapter 3 to read up on the equipment you need in order to weld safely. On top of safety equipment, of course, you also need some welding tools. Round up the items in the following list, and head to Chapter 4 for details if any them look unfamiliar.

- Angle grinder with a sanding disk
- Soapstone
- Electric ⅜-inch drill with drill bits for drilling metal
- Tape measure
- Combination square
- Hammer
- Center punch
- C-clamps
- Bar clamps
- Tool for cutting (either a demolition/reciprocating saw with a metal blade or a cutting torch)
- Stick or mig welding machine

Acquiring the right steel pieces

Perhaps the most important supplies you need for building a steel cart are, well, pieces of steel. This cart project uses mild steel. You probably won't be able to buy these specific lengths, so you'll have to buy longer or bigger pieces and cut them down. (See the following section for instructions on measuring and cutting the pieces.)

- One 20-inch piece of 1-inch-diameter steel pipe
- One 20-inch-x-¼-inch-x-1-inch flat piece
- Two 24-inch-x-¼-inch-x-1-inch flat pieces
- Two 40-inch pieces of angle iron 3⁄16 inch x 1½ inches x 1½ inches
- Two 14-inch hard rubber wheels with a bolt-on axle kit

You can use wheels of different sizes, and you can use inflatable wheels if you want. I like 14-inch wheels because they roll easily, and I'm a big fan of hard rubber wheels rather than inflatable wheels. Carts with inflatable wheels may roll a little more smoothly, but a welding shop presents lots of ways to puncture a tire, and I'd rather be spending time welding than having to air up, patch, or replace torch cart wheels.

- One 36-inch length of steel chain (1-inch links work well)
- Two pieces of ¼-inch-diameter round stock bent into *J* shapes. You can buy steel and hammer it around another piece of round stock to make a *J* shape.

If you buy this material already in a *J* shape, it's probably going to be coated with cadmium, which is toxic. If you're in that situation, use a respirator when welding those pieces.

Figure 16-2 shows you all the necessary pieces.

These pieces form the box:

- One plate of 5½-inch-x-17½-inch 12 gauge steel
- Two plates of 3½-inch-x-5½-inch 12 gauge steel
- Two plates of 3½-inch-x-17½-inch 12 gauge steel

These pieces form the base:

- One plate of 20-inch-x-10-inch 12 gauge steel
- Two pieces of 20-inch-long angle iron 1 inch x 1 inch x 3⁄16 inch
- Two pieces of 9½-inch-long angle iron 1 inch x 1 inch x 3⁄16 inch

Measuring and cutting pieces

This section is where the fun part starts. The first steps focus on making some measurements and cutting the larger pieces of metal into the sizes you need for the specific pieces I list in the preceding section.

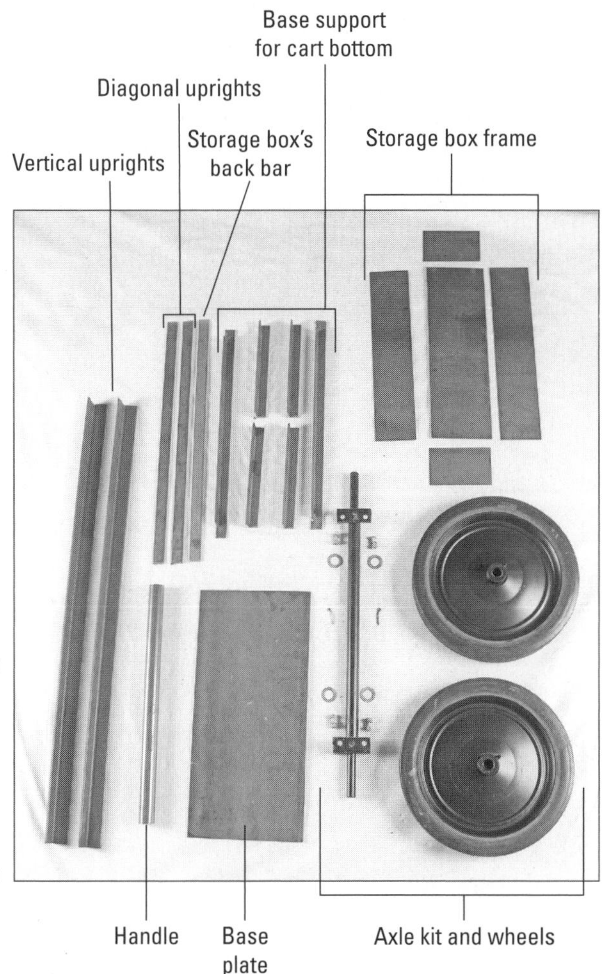

Figure 16-2: The steel pieces you need to build your torch cart.

1. **Using your tape measure and soapstone, measure and mark one piece at a time.**

   If you're cutting sections off a long piece of metal, make sure you measure one piece at a time and make all the measurements from one end. Here are some pointers to keep in mind as you're measuring the various pieces:

   • When you're making measurements to prepare for cutting, be sure to allow for the width of your saw blade or the *kerf* (waste material) caused by your torch.

- When you're cutting angle iron, draw your measurement marks on both of the "angles" so you can make a more accurate cut through the whole piece.
- For the longer pieces and plates, use a center punch to punch a hole in the measurement lines every half inch or so. That helps ensure that your lines are accurate and allows you to stay on the mark if you're cutting with a torch and the soapstone line melts away or gets obscured by waste material.
- For the parts with square edges, be sure to use your combination square.

2. **After you make all your marks, use your saw or torch to cut along the marks.**

If you have trouble cutting directly on your mark, try to cut long instead of cutting short — you can always go back later and grind down the excess if you need to, but adding material back in is problematic.

3. **When you're done making cuts, lay out all the parts like you see in Figure 16-2 earlier in the chapter.**

That may seem like an unnecessary step, but take it from a guy who has done thousands of welding projects: Making sure you have all the parts present and in the right size before you start building is a great practice. You can lay out small- to medium-sized projects on your welding table, but use the floor for projects with larger pieces.

Making the welds

Before I walk you through the steps to put together the cart, I want to mention the importance of making good tack welds. *Tack welding* allows you to hold your parts in place until you're ready to weld them, and for this project (and most others), it's essential. It allows you to focus on making good welds because you don't have to worry about holding things in place. It also helps keep *distortion* (warping of the metal) to a minimum.

Your goal is to make tack welds that are small enough to be incorporated into the finished weld. If they're too big, you may need to grind them down a little to make sure you get complete fusion on your finish welds. Always make your tack welds by using whatever welding process you plan to use to make your finish welds.

If the project you're working on has corners, make your tack welds ½ inch from those corners.

Can you tell how much of an advocate I am for tack welds? I don't even try to hide it. Now that you're also a believer (I hope), get ready to start welding. You actually tack together most of the cart before you begin making any finish welds.

Not all your steel parts are going to fit together exactly as they're designed, because your cuts will inevitably have some imperfections. Don't sweat it, though. If a couple of parts don't fit together exactly right, use your grinder to grind away the problem areas. If the pieces are still being difficult, you can force fit them by tack welding one of the pieces and then using a hammer to pound it into place. Just don't strike the hammer on any part of the piece that may damage the surface or make the part unusable. If you need to move a part more than ⅜ inch, a bottle jack may work better than a hammer.

Tacking the base and handle

You start out the project by putting together the cart's base and adding the handle. Here's how you do it:

1. **To create the base, lay the 20-inch and 9½-inch angle irons on your welding table to form a rectangle and tack the pieces together, keeping in mind that you want to make tacks ½ inch from the corners.**
2. **Tack your base plate (the 20-inch-x-10-inch piece of 12 gauge steel) on the angle iron frame and set the tacked-together base aside.**
3. **To lay out the handle, place the two 40-inch angle irons side by side on your welding table so the flanges on both pieces touch; measure 24 inches from the bottoms of the pieces and make a mark across both.**
4. **Bend the angle irons to make the two arms of the handle by using your saw or torch to cut one flange on each of the angle irons at the marks you made in Step 3; put both pieces in a vise or use your clamps and (using your gloved hands) bend both pieces to create an approximately 15-degree angle where you cut the flanges.**

   Now you should have a *V*-shaped opening in both pieces of angle iron.
5. **To patch the *V*-shaped openings in the angle irons, remove the angle irons from the vise or clamps and lay one piece on top of a piece of scrap steel (with the same thickness) so the scrap is underneath the *V*; mark the shape of the *V* on the scrap piece with your soapstone.**
6. **Cut out the scrap *V* with your torch or saw so that you have a small *V*-shaped steel chip to patch the *V*-shaped opening in the angle iron.**
7. **Fit the chip into the *V* on the angle iron (you may need to grind it smooth on both sides to get a good fit) and weld it so that the chip becomes part of the angle iron.**

Be sure to weld it on both sides and then smooth out the weld joints with your grinder.

8. **Repeat Steps 5 through 7 to patch the other *V*-shaped opening on the other angle iron.**
9. **Weld your two bent angle irons to the base one at a time.**

   The angle irons should fit on the outside of the long sides of the base, going all the way to the floor. The flanges on the angle irons should face inward so they can receive the pipe handle at the top. Use your combination square to make sure the pieces are truly perpendicular to the base and parallel to each other.

   By this point, you have your base tacked together and the two arms for your handle tacked onto the base. That's no small feat, so give yourself a quick pat on the back.

10. **To complete the handle, tack your 20-inch section of steel pipe to the free ends of the angle iron arms.**
11. **To form a brace that runs from the angle iron arms down to the other end of the base, position one of the 24-inch-x-¼-inch-x-1-inch flat pieces on one side at the end of the base opposite from where you attached the angle iron arms.**

    Tilt the piece so it runs from the base up to the angle iron arms. Then use a soapstone to mark the angle of the base onto the piece, and cut that angle (with a reciprocating saw, torch, or even a grinder).

12. **Repeat Step 11 for the other piece on the other side of the base.**
13. **For even more stability, tack the piece of 20-inch-x-¼-inch-x-1-inch steel flat across the back of the cart where the angle begins, making sure the piece is level.**

    How's the cart looking now? Hopefully, it looks something like Figure 16-3.

Building the box and finalizing the welds

Now you're ready to build the box, using the five pieces of 12 gauge steel (five because the box doesn't have a top). Just follow these steps:

1. **Tack the 5½-inch-x-17½-inch piece to one of the 3½-inch-x-17½-inch pieces so that they make a 90-degree angle at the long side.**

   Two or three tacks per side should be enough. Remember that you shouldn't tack in the corners.

2. **Repeat Step 2 to add the other 3½-inch-x-17½-inch piece to the setup.**
3. **Fit one of the 3½-inch-x-5½-inch plates so it's on the end and forms 90-degree angles with the larger plates you've already tacked together; make sure it looks square and flush, tack it on, and repeat for the other 3½-inch-x-5½-inch piece on the other end of the box.**

Figure 16-3: The partially welded torch cart.

4. **Now that the box is square, true, and flush, put small ½-inch welds every two inches on the box's inside joints.**

   Don't overweld or the sides of the box may warp.

5. **To attach the box to the base, tack one of the long sides (just below the top edge) right below the bar you welded across the back of the angle iron in the preceding section.**

   If you have any doubts about the placement of the box, refer to Figure 16-1 earlier in the chapter to see where the box goes on the cart. (Make sure you locate it so it'll be on the back of the finished cart!).

6. **Make sure the box is level and then weld two small (½-inch) tacks on the top and bottom.**

   That should be enough to hold it for the time being.

7. **To attach the *J*-shaped hooks that hold the chain that secures your gas cylinders, measure 26 inches up from the base of the cart and make a mark with your soapstone on the outside of the two vertical angle irons; tack one of your *J*-shaped hooks on each side, making sure they're both at the same height.**
8. **Go back and weld all the tack welds you made, working from the base of the cart toward the top.**

Adding the wheels

You're now the proud owner of a torch sled (see Figure 16-4). But that doesn't do you much good, of course — you need a torch cart, and a cart has to have wheels.

Figure 16-4: What your torch cart should look like before adding wheels.

The following steps show you how to add those all-important elements to your cart.

1. **Put a few pieces of ⅛-inch scrap metal underneath the base of the cart so it's propped up off the floor.**
2. **Roll the wheels up against the back of the cart and line up the mounting brackets so that they're flush with the angle iron frame.**
3. **Stick a sharpened soapstone through the holes in the mounting brackets so that you make marks on the angle iron.**

   Make sure to make marks for all the holes — you may have several.
4. **Use a hammer and your center punch to punch holes in the center of the soapstone marks.**
5. **Using those holes as pilot holes, drill holes that are appropriate for the size of the bolts that came in your wheel kit.**

   The kit should include a set of instructions that tell you what size bit to use.
6. **Bolt the wheels onto the frame.**
7. **Load up a couple of gas cylinders into the cart and secure them by running the chain across them and attaching each end of the chain to one of the *J*-shaped hooks; tilt the cart back and push it around to make sure it rolls smoothly.**

Your cart should now look like the one in Figure 16-1 — without the gas cylinder, of course.

Checking your welds

After you've made all the welds and tested the cart, take some time to look carefully over each weld. You may want to spend a few minutes using your grinder to make sure the welds are smooth.

Make sure you grind down the welds on the outside edges of the cart, especially in the handle area. You don't want jagged welds snagging your equipment or clothes (let alone your hands).

After you finish grinding, you may want to go over some of the welds with a wire brush to really smooth them out and clean them up. Then you can use a rag and a cleaning solvent (any cleaner that cuts grease will work) to wipe down all the steel to get it ready for painting. Believe me, if you have any really jagged welds left, you'll find out really quickly when you run over them with a rag.

Picking out your paint

If you opt to paint your torch cart, keep these guidelines in mind when you're choosing paint.

- Any color is fine, but use a flat paint that so it doesn't reflect the ultraviolet rays that come from your arc welding projects.
- Be sure to pick a paint designed for use on steel.
- Look for paints that include a rust preventative; it's not required, but I highly recommend it.
- Paint fumes can be hazardous, so be sure to paint where you have adequate ventilation and follow the instructions that come with your paint.

Fabricating Your Own Portable Welding Table

The portable welding table (also known as a *parts cart*) makes a terrific project for novice welders. You may already have a good, sturdy, stationary welding table, but if you continue to practice welding, you may eventually want to be able to move around a little, and a portable welding table adds some mobility to your welding activities.

If you've built up some welding skills and you want to put them to good use, this chapter is for you. Read on to find out how you can fabricate a great portable welding table.

This portable welding table is 2 feet x 3 feet — good-sized, but not enormous. I've found that this size allows you to hold and move around a lot of tools and materials but isn't so big that you really have to put your back into it to push it, or worry about it banging into things all the time. You can fit it through any doorway, too, which is always really helpful. (Nothing is worse than slamming into a door jamb and damaging the jamb, your cart, or both!)

Before you get started, take a good long look at the finished portable welding table in Figure 16-5 to get an idea of where you're headed.

Figure 16-5: A completed portable welding table.

Rounding up your tools

The most important equipment for this (and any) welding project is your safety and protective equipment. Make sure you have the appropriate gloves, safety goggles, and clothing, plus a good welding helmet that protects your eyes from the rays put out by the welding arc. That list only scratches the safety surface, so be sure to check out Chapter 3 to read up on all the welding safety details.

When you've got your safety equipment in order, here's a list of the specific tools you need to build this project.

- Angle grinder with grinding and sanding disks
- Soapstone
- Electric ⅜-inch drill with drill bits for drilling metal
- Tape measure
- Combination square
- Steel file
- Reciprocating or cutoff saw
- Stick or mig welding machine

Picking out the parts

This project is pretty basic, so you don't have to spend a lot of money buying a ton of parts you need. Here's the rundown.

- Two 24-inch-x-36-inch pieces of 3⁄16-inch-thick flat steel
- Four 32-inch pieces of 3⁄16-inch-x-1½-inch-x-1½-inch angle iron
- One 3-inch piece of 1-inch-diameter steel pipe
- One 2-inch-x-2-inch piece of 3⁄16-inch-thick steel plate
- Four 4- to 10-inch bricks, cinder blocks, or other sturdy pieces of material to use as spacers
- Four castors with matching nuts and threaded shafts
- Four washers that fit the castors

When you're trying to decide which castors to buy, keep in mind that the larger the wheels, the more easily they roll. I'm a big fan of the castors with 4-inch wheels for this project — they're great if you're of average height. Wheels of that size give you a working surface on the top of the cart that's about 38 inches off the floor. I also recommend that two of the four castors be the kind you can lock into position.

If you can't buy steel pieces in these sizes, you may need to buy larger sizes and cut them down with a saw or torch. You can see all the parts you need for the portable welding table laid out in Figure 16-6.

Assembling the pieces

Start by putting on all your personal protective gear, and make sure your welder is set for working on 3⁄16-inch steel. Then follow these steps.

1. **Lay one of your 24-inch-x-36-inch flat steel sheets on the floor and lay your welding machine's ground clamp on top of the sheet.**

   Of course, I'm talking about the concrete floor of your welding shop, not the shag carpet floor of your living room.

2. **Hold one of the 32-inch angle irons vertically so that it's perpendicular with the large steel sheet; place the end of the angle iron in the corner of the sheet so that the corner of the angle iron is flush with the corner of the sheet.**

3. **Make tack welds that join each of the two flanges of the angle iron to the sheet.**

   Make sure that it's square! *Tack welds* are small welds that hold pieces together until you're ready to finalize the welds. You then incorporate the tack welds into your final weld.

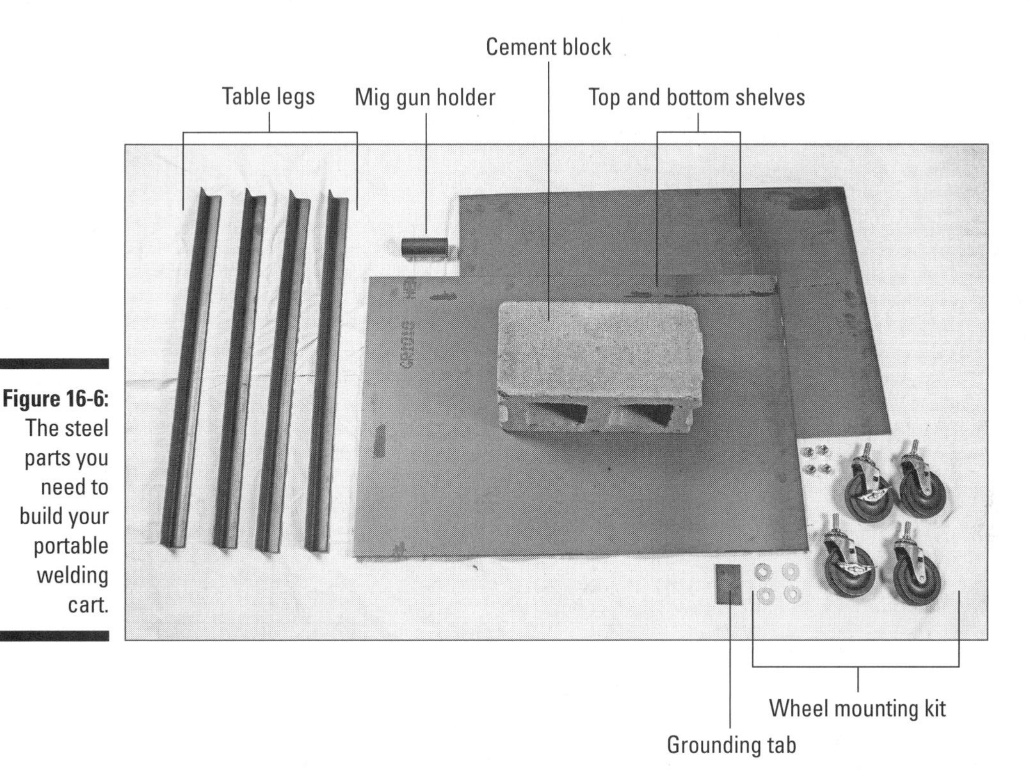

Figure 16-6: The steel parts you need to build your portable welding cart.

4. **Repeat Steps 2 and 3 for the other three corners to finish attaching the table legs; after the legs are in place, go back and weld over the tack welds to fully secure the legs to the plate.**
5. **Flip the table so that it's standing on its legs and slide the washers one at a time into the corners of the angle iron legs.**
6. **Tack the washers to the angle irons in that position.**

   Make sure to tack each washer in at least two places for strength and stability. If positioning the washers for welding seems difficult, check out Figure 16-7.
7. **Put the spacers underneath each of the legs so that the table is elevated several inches off the floor (see Figure 16-8).**
8. **Maneuver the other flat steel sheet between the angle iron legs and rest it on top of your spacers.**

   You may need to tilt the sheet to get it between the angle irons, and you need to make sure that each of its four corners is resting on a spacer. Check that both steel sheets (the one you've already welded and the one you haven't) are level.

Figure 16-7: Position for welding the washers to the pieces of angle iron.

Figure 16-8: The partially constructed table lifted off the floor with spacers.

9. **Tack weld the bottom sheet to the angle irons in at least two places per corner and flip the table over again.**

10. **Go through and make the final welds on all the joints.**

    The welds are going to be at the corners, of course, so use an alternating pattern as you weld: Make the welds at one corner, move diagonally to the next corner, slide over to the next corner, and then move diagonally again to the last corner.)

11. **Weld the 2-inch-x-2-inch plate to the bottom of the tabletop at a 90-degree angle so you can attach your ground clamp to it later.**

12. **Weld the 3-inch section of pipe onto one of the angle irons about 8 inches down from the tabletop.**

    You can use that little piece of pipe to hold your mig welding gun (like a holster) when you're not using it.

How does your portable welding table look? It should look like the one in Figure 16-5 earlier in the chapter (minus the paint job, of course; head to "Choosing your paint" later in the chapter for more on painting the cart).

Ensuring smooth edges

Because you're going to be using your portable welding table (hopefully all the time) and working on and around it, you want to make sure that the welds on the table are smooth and don't catch things that brush by. To make sure your welds are smooth, take a few minutes to examine each one carefully. If it's not smooth to the touch, put a flappy sanding disc on your grinder and sand all the edges smooth. You may also want to use your steel file to smooth down the inside of the little section of pipe and the edges and corners of the grounding plate. (You probably can't fit your grinder in those tight spots safely, so the file is the best bet.)

When you use your grinder, make sure it's properly grounded and is in working order. Don't let the sparks from your grinder go toward any person or any flammable materials or equipment in your shop. Also be sure to hold on tight to the grinder, because it can catch and jerk violently when you're grinding grooves. If you're using a bench grinder and can't hold onto the piece with both hands, use vise grips.

Choosing your paint

I recommend that you paint your portable welding table. To help the table withstand the elements in your welding shop, find the type of paint used to paint charcoal grills in a flat finish. That way, the paint will be able to withstand temperatures up to about 1,100 degrees Fahrenheit, and it won't reflect the light from the dangerous ultraviolet rays of your welding arc.

Don't paint the top of your portable welding table. Don't paint the grounding plate, either.

Putting on the wheels

After the paint is dry, you're ready to attach the castors. Just follow these steps.

1. **Flip the table over so the tabletop is on the floor and the legs are sticking up in the air; bolt on all four castors, making sure they're attached securely.**

   Castors can loosen up over time if you roll your table around a lot.

2. **Flip the table over and make sure all four of your wheels are touching the ground so that the table doesn't rock.**

   If the castors don't all touch the floor, first make sure that the table is on level ground — an inconsistency in the floor can sometimes make one of the wheels seem like it's off when it really isn't. If that doesn't fix the problem, find out which castor is up off the floor. Mark that wheel, and measure how far it's away from the floor. Then flip the table and put enough washers in the castor assembly to get the wheel down so it's level with the others.

3. **Enjoy your new portable welding table!**

Chapter 17

Constructing a Campfire Grill

In This Chapter

- Using your welding skills to cook up a grill
- Discovering how to season your grill

If you've ever been camping in an area where grills or other built-in cooking equipment is available, going back to choking down hot dogs that you've burned on a stick over an open campfire can be tough. If you're going to be out in the woods and enjoying nature, shouldn't you be able to enjoy a good meal while you do it? I certainly think so, and it's in that spirit that I present this welding project: building a campfire grill. I love this project because it lets you practice your welding skills, and when it's all said and done, you have a useful item that makes braving the great outdoors with family and friends a little more enjoyable.

Fabricating a Campfire Grill

I really like this campfire grill because it's simple and extremely functional. It's very easy to assemble and disassemble, so you can spend more time enjoying your camping trip and cooking good food and less time putting grill parts together. Adjusting the height of the grill (and therefore the cooking temperature) is simple, and it's also easy to move the cooking surface onto and out of the heat. The grill is also really durable; it has very few parts that can break, and you can easily replace any piece you happen to lose on your way out of the woods.

I think getting a good look at something before you try to build it is always helpful, so take a minute to review the picture of the finished campfire grill in Figure 17-1.

Getting your tools in order

Here's a list of the tools and equipment you need for this project. Remember that you're making a campfire grill, not putting together a satellite, so the amount and variety of items you need is pretty limited.

Figure 17-1: A finished campfire grill.

- Hacksaw or band saw with blades to cut steel
- Angle grinder with disks for grinding steel
- Half-moon file
- Tape measure
- Soapstone
- Grease remover
- Stick or mig welding machine and electrodes or electrode wire appropriate for the pieces listed in the following section

Obtaining the proper steel pieces

Use mild steel for this campfire grill. (Check out Chapter 2 for more on mild steel.) I use tubing because one tube fits inside the other, where pipes don't. Here's a list of the pieces you need.

You likely won't find these steel pieces available in these sizes, so you'll probably need to buy larger sections and cut them down to size, a task I discuss further in the following section.

- Three 22-inch pieces of 1-inch-diameter pipe-size tubing
- Two 10-inch pieces of $\frac{1}{4}$-inch-round stock
- Two 2-inch pieces of $\frac{1}{4}$-inch-diameter pipe-size tubing
- Two 4-inch pieces of 1$\frac{1}{4}$-inch-diameter pipe-size tubing
- One 10-inch piece of 1$\frac{1}{4}$-inch-diameter pipe-size tubing
- One 14-inch-x-22-inch rectangular grill grate
- One 74-inch piece of 1-inch angle iron

In addition to these materials, you also need a grill grate. You can either buy a premade 14-inch-x-22-inch rectangular grate or buy the steel pieces to make one. If you choose to make the grate yourself, get twenty-seven 14-inch pieces of $\frac{5}{16}$-inch-diameter round stock, plus a 22-inch piece of that stock for spacing. You also need one 74-inch piece of 1-inch angle iron.

To get a feel for what all these pieces look like in one place, take a look at Figure 17-2. If you're building your own grate, you can see the pieces in Figure 17-3.

If you're working on this (or any) project and your welding table isn't big enough to hold pieces that you need to weld, take caution before moving the project off onto your concrete floor and welding it. Concrete contains moisture that rapidly changes to steam when it comes in contact with a welding arc or flame, and that can cause the concrete to explode violently. If you have to weld or cut on concrete, be sure to protect the floor (and yourself) by laying down a piece of scrap sheet metal first.

Cutting the steel pieces to length

Unless your hardware store or welding supply shop happens to have a lot of unusually sized pieces in stock, you'll most likely need to buy larger pieces of steel and simply cut them down to the lengths listed in the materials list in the preceding section. That shouldn't be too difficult; just mark everything with your tape measure and soapstone and then cut it down with your hacksaw or band saw.

After you cut the three 22-inch pieces of 1-inch-diameter tubing, you need to modify one end of two of those pieces so that when you place them perpendicular to the other piece, they sit on top like a saddle. Use your grinder or a half-moon file to accomplish that effect; just file or grind two of the ends of two different pieces until they're shaped like an arch.

If you choose to build the grill grate instead of buying it, you need to get those pieces ready as well.

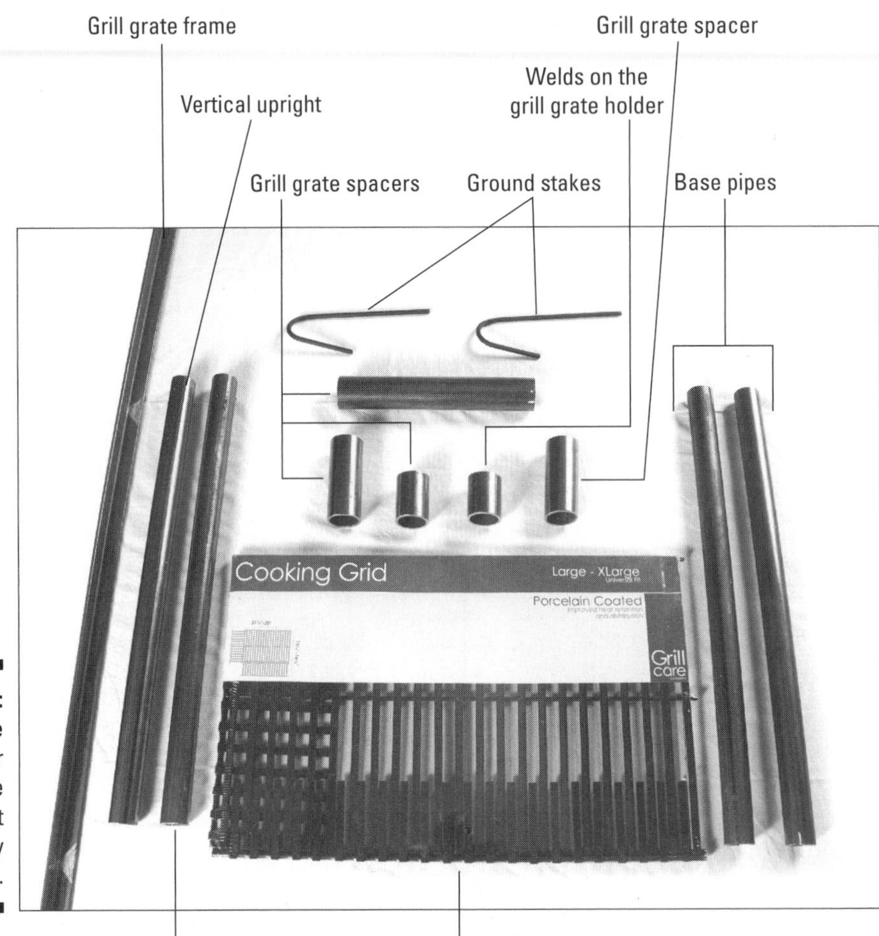

Figure 17-2: All the pieces for the campfire grill laid out and ready to use.

Cut the angle iron in four pieces; they'll form a 14-inch-x-22-inch rectangular frame around your grate with ¼ inch of extra length on both sides of the frame. (That extra length allows for some expansion due to the heat of the campfire.) Then cut the 5⁄16-inch-diameter round stock so you have twenty-seven 14-inch pieces.

After you do all your cutting, go over your cut edges with a file to make sure they don't have any rough or jagged spots. I also recommend wiping down all the parts with grease remover.

Cut the angle iron in four pieces; they'll form a 14-inch-x-22-inch rectangular frame around your grate with ¼ inch of extra length on both sides of the frame. (That extra length allows for some expansion due to the heat of the campfire.)

Figure 17-3: Pieces for the grill grate.

After you do all your cutting, go over your cut edges with a file to make sure they don't have any rough or jagged spots. I also recommend wiping down all the parts with grease remover.

Welding the grill

After you have all your pieces cut and filed down to ensure no jagged edges, you're ready to start making some welds. You can start by tacking together the pieces.

As with most projects, welding the pieces that make up this campfire grill require that you *tack weld* everything together first before making the finished welds. Tacking is a great way to restrict movement of the pieces when you're making a finished weld, which is key if you want to end up with high-quality welds. Your finish welds then incorporate your tack welds.

It's also always a good idea to weld opposite corners first; that helps to prevent warping.

To build your grill, just follow these steps.

1. **Lay the three 22-inch pieces of tubing flat on your welding table so that the two with modified ends are saddled on top of the other piece, creating a *U*-shaped frame; tack those saddled joints together.**
2. **Flip over that *U*-shaped frame and make a finished weld on that side of the saddled joint.**
3. **Flip the *U*-shaped frame back over and make a finished weld on that side.**
4. **To weld the grill grate (if you're building your own), lay out your angle iron pieces so that the two 14-inch pieces are on one side and the two 22-inch pieces are on the other side to form a 22-inch-x-14-inch rectangle; make sure the corners are square and then tack weld them together.**
5. **Lay the twenty-seven pieces of $\frac{5}{16}$-inch round stock inside the L of your angle iron frame and push them all to one side; tack weld every other piece of round stock to the angle iron frame on both sides and remove the pieces of round stock that you didn't tack weld.**

   Half of your cooking surface should be close to finished right now.
6. **Put the round stock you removed in Step 5 back into the frame where there are gaps between the round stock you tack welded and repeat the tacking and removing from Step 5; continue replacing the removed pieces and tacking in this manner until you've tack welded all the round stock into the frame.**

   Just space the placement by eye.
7. **Weld the round stock on the top of the grill grate, trying to seal the round stock to the angle iron, and weld the four corners of your frame.**

   Figure 17-4 shows the finished grate.
8. **Tack one of the 2-inch pieces of $1\frac{1}{4}$-inch tubing on the inside of the bottom piece of the *U*-shaped frame, making sure to tilt it back 5 to 10 degrees and center it; after you're comfortable with the angle, finish the weld.**

   The 5- to 10-degree angle you add allows your grill grate to stay level when you put weight put on it. By this point, your project should look like the one in Figure 17-5.
9. **Bend your two pieces of 10-inch round stock into *U* shapes and set them aside.**

   Those pieces stake down the base of the grill so that it doesn't move when you swivel the grill grate on and off the fire.

Figure 17-4: The assembled grill grate.

Figure 17-5: Partially constructed campfire grill.

10. **Tack your second 2-inch piece of 1¼-inch diameter tubing onto one of the long sides of the grill grate frame about halfway down the 22-inch piece, ensuring that it's square and centered, and then go back over it to finish the weld.**

You should still have two pieces of 4-inch long pipe and one piece of 10-inch long pipe. Those will allow you to make height adjustments for the grill grate so you can set it at different heights above the fire. You should be able to achieve up to 18 inches of height adjustment.

11. **Finish any tack welds that you haven't already incorporated.**
12. **Assemble all the parts so your grill looks like the example in Figure 17-1.**

Picking out your paint

You're going to be using your grill over an open, roaring campfire, so you know it's going to be exposed to some high levels of heat. Because of that, not just any paint is appropriate. Make sure you pick a paint tested to withstand high temperatures; the paint sections in some hardware stores offer paints that are specially made for charcoal grills, gas grills, and fire pits. That's the kind of paint you want to get. Read the label before using the paint to be sure it can provide what you need, and remember that two thin coats are better than one.

Don't paint any surface of the grill on which food will be cooked (including your grill grate, of course).

Seasoning the cooking surface

After you have your campfire grill assembled and painted, you may want to take the time to season the cooking surface. Doing so keeps food from sticking to the grate and also imparts a little flavor. You can season the surface by using an oxyacetylene torch; for info on setting up one of those torches, flip to Chapter 13.

1. **Make sure you're wearing your protective gear, including welding gloves.**
2. **Use the torch to heat up the cooking surface to 400 to 500 degrees Fahrenheit.**
3. **Put cooking oil on a clean rag (one that you never need to use again), and wipe it on the entire cooking surface.**

   Don't be surprised if it smokes quite a lot as you're working.
4. **Allow the metal to cool down completely before you do anything else with it.**

Chapter 18

Fixin' to Fix Things: Analyzing and Planning

In This Chapter

- Figuring out whether repairing or replacing makes more sense
- Coming up with a strategy for your welding repairs
- Preparing for repair projects
- Taking a look at cracks

A lot of people get into welding because they want to be able to repair metal items. Whether it's a metal tool, a piece on your car, or an important component on some farm machinery, metal objects — despite their unmatched toughness — do occasionally break. When that happens, replacing them can be an expensive or inconvenient endeavor, and sometimes replacement isn't an option at all because a part may be no longer available.

This chapter is all about using welding techniques to make repairs. But don't expect me to jump right into instruction on how to run a weld bead to fix a broken joint. If making repairs were that quick and easy, the market for new metal objects would be in the tank, and you know that's not the case. No, repairs can be complicated animals that require a lot of analyzing and planning, and if you don't get those steps right, the chances of your ending up with a fully repaired piece are slim to none. To help with that foundation, I also show you how to use your judgment (and plenty of welding know-how) to make sound decisions about repairing metal parts with welding, and then execute those plans successfully.

Determining Whether Something Is Fixable

New welders often ask me about the first step in the repair process — whether they should start by figuring out how to set the wire feed speed on their mig welding machines, or maybe by cleaning up the broken metal piece with an angle grinder. I usually let them keep guessing for a minute or two before I tell them the correct first step for any repair job: figuring out why the piece broke in the first place. Most people look at me a little funny, because they either didn't think of that or felt it was too obvious to bring up. But it's really a critical step, and if you don't take that step, you can easily end up wasting your time as you try to make a weld (or several) for repair purposes.

For example, say you buy a piece of equipment designed to bear a 100-pound load. Eventually you start using it more and more, and the scale of your projects continues to go up and up. Soon you need your equipment to be able to bear a load of 200 pounds, and then 300. The equipment — if it's made well — may hold out for a little while, but if you keep asking it to do three times what it's designed to do, you're headed for disaster. Sure enough, eventually an important metal part on the equipment breaks.

That's when you may start thinking about your options for using your welding skills to repair the broken piece. But wait! Stop right there. What you really need to do is take a minute to consider what happened to the equipment in the first place. It didn't fail. You did! You knew it wasn't rated for 300-pound loads, but you kept surpassing the limits anyway. In this case, welding the broken metal piece back together so you can go back to overloading the equipment isn't the way to go. You don't need to fix your 100-pound-rated equipment. You need to get your wallet and go buy a piece of equipment that can handle a 300-pound load.

But if, after careful consideration, you determine that irresponsible use isn't why your piece broke, your next step is to decide whether repairing the item yourself is feasible or whether you need to buy a replacement instead. Ask yourself the following questions:

- **Will your welding repair cost more than buying a new piece?** I'm not saying your entire decision has to be about dollars and cents — if the tool in question came from your great-grandpa and continuing to use it is sentimentally important to you, no amount of monetary savings a replacement may provide will outweigh the satisfaction of using something that means a lot to you.

But most repair decisions aren't that emotional; in most cases, you can easily weigh whether the cost of materials and your time (which is also valuable!) are going to exceed the amount you'd spend if you went out and bought a new piece. If the repair is going to take $100 in filler rods and buying a new piece would cost only $85, you're probably better off just buying the replacement.

- **Will your welding repair work compromise other pieces of the tool or equipment you want to fix?** Repair welding is still welding, and it still produces a huge amount of heat and some other dangerous, damaging forces. A repair that does more harm than good isn't much of a repair at all. If a metal bracket under the driver's seat of your car breaks and causes your commute to feel like an amusement park ride, you may think, "Hey, I can weld that back together no problem."

  But before you grab your welding helmet and turn on your welding machine, take a minute to really think about what you may damage when you're trying to weld the bracket back together. If you discover that the welding repair will likely severely damage the $215 power seat motor situated right by the bracket, you may very well end up down at the dealership ordering a new power seat motor. Then you're out not only the $215 (plus the labor costs for having the motor installed) but also the time you spent going back and forth to the dealership four times (because you know they're not going to have the part you need when you first go down there).

- **Will the repair weld be stronger than the original piece?** This question is very important. Your goal for every repair weld is a finished piece or joint that's stronger than the one that broke in the first place. If that's not possible, I encourage you to really consider your options for replacement.

Planning a Repair Strategy

Say you're faced with a potential welding repair situation and you work your way through the steps described in the preceding section. You figure out that using welding to make a repair is indeed the best route, and you want to get started. Not so fast! You need to first come up with a repair strategy that ensures success, or at least gives you the best possible chance at succeeding. A good repair strategy involves identifying the metal you're dealing with, choosing the best welding technique for the job, and making and following the plan for the repair. The following sections give you the details on doing just that.

Identifying the metal and what it means for the repair

You can't even hope to make a good welding repair unless you know the kind of metal you're working with. Here are a few options for figuring out the type of metal involved in your repair.

Before you try to identify a metal, always take a minute to clean it off first. That way, you can see the true color of the metal.

- **Check the literature.** If you're going to be repairing a tool or piece of equipment that came with an instruction manual, specifications sheet, Material Safety Data Sheet (MSDS), or similar documentation, it may list the type of metal somewhere in those details. Check them over before you go any further.
- **Do a quick color test.** This test is extremely easy to conduct. Just look at the piece of metal. Is it gray? If so, you may very well be looking at a piece of cold rolled steel or cast iron. Is it shiny and/or light colored? If so, you may have aluminum or magnesium on your hands. The color of a metal isn't a completely foolproof way to identify the metal, but it's a really good starting point. To help in your efforts to identify a metal by its color, consult Table 18-1.

Table 18-1 Metal Color Identification Chart

Metal	*Color of Unfinished, Unbroken Surface*	*Color and Structure of Newly Fractured Surface*	*Color of Freshly Filed Surface*
White cast iron	Dull gray	Silvery white; crystalline	Silvery white
Gray cast iron	Dull gray	Dark gray; crystalline	Light silvery gray
Malleable iron	Dull gray	Dark gray; finely crystalline	Light silvery gray
Wrought iron	Light gray	Bright gray	Light silvery gray
Low carbon and cast steel	Dark gray	Bright gray	Bright silvery gray
High carbon steel	Dark gray	Light gray	Bright silvery gray
Stainless steel	Dark gray	Medium gray	Bright silvery gray
Copper	Reddish brown to green	Bright red	Bright copper color

Metal	*Color of Unfinished, Unbroken Surface*	*Color and Structure of Newly Fractured Surface*	*Color of Freshly Filed Surface*
Brass and bronze	Reddish yellow, yellow-green, or brown	Red to yellow	Reddish yellow to yellowish white
Aluminum	Light gray	White; finely crystalline	White
Monel metal	Dark gray	Light gray	Light gray
Nickel	Dark gray	Off-white	Bright silvery white
Lead	White to gray	Light gray; crystalline	White

- **Check for magnetism**. Get a magnet (you may have one in your shop or maybe on your refrigerator) and see whether it's attracted to the metal in question. If it is, you're dealing with a *ferrous* (iron-containing) metal. If not, the metal is *nonferrous* (iron-free).
- **Do a spark test.** This one is a little more involved than the color or magnet test, but it's still extremely easy. Using your angle grinder (and your safety gear, of course), grind a little piece of the metal. Does it spark? If not, the metal doesn't contain iron. If it does spark, you may be able to identify it by using the spark test chart in Chapter 15.

Sparks can travel up to 35 feet and still remain hot enough to catch something on fire, so be careful when you're doing a spark test and make sure no flammable materials are nearby.

- **Try a metal identification kit**. If you're not having any luck with any of the preceding tests, you can always buy a metal identification kit at your welding supply store or online. These kits use chemicals to identify different metals. They're not too expensive, and they're generally really accurate.

If you think you may be working with an *alloy* (combination of metals), it may be worth using a metal identification kit. They're great for determining specific alloys, and that's something that can be hard to do with other methods.

After you figure out what kind of metal you're going to be repairing, take a minute to think about what you know about that metal and how that informs the rest of the process. The following bullets highlight a few notes about repairing common metals that you may want to keep handy (or at least keep in mind).

- **Aluminum:** Aluminum has an oxide coating on it, so you have to do away with that coating before you can effectively weld to repair aluminum. Just brush the surface near the weld area vigorously with a stainless steel wire brush or even some rough-grit sandpaper. Also remember that aluminum doesn't change color when it's heated, so don't expect it to turn orange or red like steel does when you heat it up. In other words the piece may be hotter than it looks, so don't touch it.

If you're going to be making welding repairs on aluminum, you may want to get some information on welding the metal and its alloys from one of the major aluminum producers. All of them provide welding manuals and charts for welding their products. You can find their information online; if the details you want aren't listed on their company Web sites, give them a call and ask where you can find it. Some pieces of aluminum have an identification number of some kind stamped on them; it's a good idea to find that number, write it down, and keep it handy when you're researching details of how to weld the aluminum piece.

- **Copper:** When you need to repair a piece of copper, take into account that it transfers the heat from welding faster than just about every other commercial metal. It doesn't change color when heated, so you can easily end up with a superheated piece of copper that looks like it's at room temperature.

For general information on copper and how to weld it, check out `www.copper.org`, the Web site of the Copper Development Association (CDA). The CDA established the designation system for copper used in the United States, and it's a good resource for copper data and info. Its bulletins are particularly helpful.

- **Cast iron:** Gray cast iron is one of the least expensive metals for making parts, so you're likely to run into it if you're making a lot of repair welds. It's popular in automotive applications such as engine blocks, engine heads, rear-end housings, brake drums, and more.

If you're challenged with welding a piece of gray cast iron that's going to be heated and cooled repeatedly, think hard about replacing the piece. Gray cast iron exposed to that kind of environment often breaks down, and the service life of your repair may not be as long as you hope.

To see an example of a welding repair done on cast iron, check out Figure 18-1.

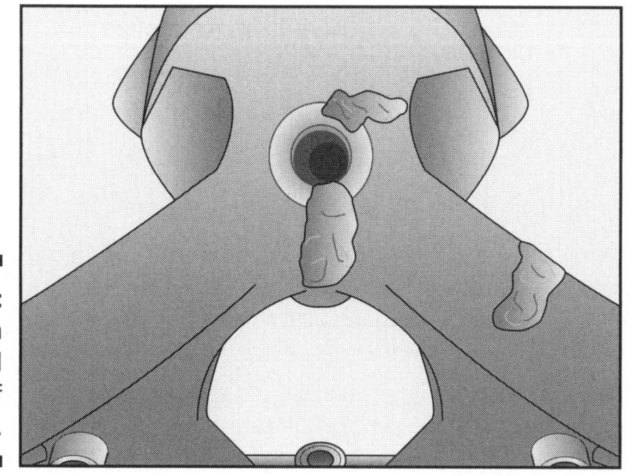

Figure 18-1: Welds in a repaired piece of cast iron.

Hardfacing

Sometimes when metal pieces are welded for repair purposes, another welding technique called *hardfacing* is used after the fact to help produce a more resilient end result. It kind of sounds like some kind of hockey penalty, but it's really a process that involves overlaying a strong metal on the surface of a weaker one. Hardfacing can help make a piece of metal resistant to damaging forces like abrasion and corrosion, and it works for building up metal surfaces that have been worn out.

Some of the most commonly used metals for hardfacing are iron, nickel, and copper; you deposit them by using an arc welding or oxyfuel welding process. If you want to try hardfacing, read up about it on the American Welding Society's Web site `www.aws.org` (you can find a great article about hardfacing in the July 2006 issue of its *Welding Journal*). You can also ask the folks at your welding supply store about hardfacing electrodes and the availability of courses that teach hardfacing techniques in your area.

Deciding which welding process to use for your repair

Here's a list of things to consider when you're trying to figure out which welding process to use for your welding repair project. I focus on the major arc welding techniques because those are the most commonly used on repair jobs.

- **The type of metal you're working on:** Certain types of welding work better on specific kinds of metal. Here's a rundown of the options for some of the most common metals.
    - **Aluminum:** Tig welding is always a good option for repairing aluminum. It's the cleanest of the welding processes and can produce some really sharp, strong welds. It works well for thinner pieces of aluminum and also on aluminum alloys, which have become pretty common. Mig welding is popular for aluminum repairs too, especially if you're working on a thicker piece. It's quite a bit faster than tig, which can be hugely helpful if you're pressed for time. Stick welding aluminum is possible, but it's not the best. Use the technique for your aluminum repair projects as a last resort.
    - **Copper:** You can use almost any welding process on copper. Soldering is a popular choice when it comes to joining copper pipe, and brazing is also widely used to join copper with other metals. (See Chapter 13 for more on these options.) And of course, tig and mig welding are great options as well.
    - **Cast iron:** Stick welding is a common choice for repairing cast-iron pieces. Brazing is another popular option for making repair welds on cast iron because you can use it to join cast iron and steel. Surprisingly, you can also use mig welding to join carbon steel and cast iron.
- **The equipment you have on hand:** If all you have is a mig welding machine, you obviously need to focus your repair plan on a weld you can make with a mig machine. Spending several hundred dollars on stick welding equipment, for example, to fix the handle on your garden hoe doesn't make much sense.
- **How much time you have to make the repair:** If you're pressed for time, you probably want to opt for mig welding if that's an option.
- **Weld aesthetic:** Think about how visible the weld's going to be and how important you feel a good-looking weld is. For example, if you're going to be welding to repair part of the body on your mint-condition 1932 Ford Deuce Roadster, you probably want to opt for the clean precision of tig welding over the messier, slightly less refined stick welding option.

- **Your skill level:** Are you pretty handy with a mig welding gun but seem to have two left hands when you work with stick welding? If so, use mig for your repair project (at least until you've had time to practice stick welding a little more.) And even if you've picked up a particular welding process pretty quickly, some repairs may still be outside your level of expertise, especially as you're beginning.
- **Condition of the base metal:** Is the metal you want to repair rusty, oily, or just in bad shape? If so, go for stick welding.
- **Strength of the base metal and surrounding pieces:** If the base metal you're repairing is strong and resilient, your options for a good welding repair technique are many. However, if the metal is weak and has a low melting point, or if the surrounding pieces have some of those qualities, you may want to try a lower impact welding process, such as a non-arc welding technique such as soldering.

Making and following your plan

After you've identified the type of metal you're going to be repairing and which welding method you want to use (see the preceding sections), you're not ready to weld just yet. First, you need to really think your way through the actual weld and make a plan that gives you the best chance at succeeding with your repair. Keep the following considerations in mind.

- **What were the mechanical pressures that caused the piece to fail?** If you can identify those pressures and make your repair weld in such a way to alleviate some of their negative effects, that's a real win for you.
- **How can you improve the piece so that it doesn't fail again?** Can you add another brace? Extend the piece a little so it has a larger surface area? You can improve the strength or flexibility of a metal piece in a lot of ways when you're repairing it with a weld.
- **Was the old product overdesigned?** This question is one that new welders may not think of. There's a good chance that the original piece was too complicated, and you may be able to simplify it with your repair weld. Keep your eyes open for those opportunities. A simple weld is usually better than a complicated one.

After you've come up with the answers to those questions, consider the joint. Here are some important things to keep in mind when you're making choices about the kind of joint to use in your repair.

- **Design your joint so it can bear the weight required by the piece with the least amount of filler rod possible.** You don't want to go too light on your use of filler rod, of course, but you definitely don't want to use too much because it's a waste of material. It also requires quite a bit more heat to properly melt a filler rod that's too large for the job, and that heat isn't free, of course.

- **Don't plan for a big joint when a smaller one will do.**
- **If you're working with a thick plate, plan to groove or bevel both sides with a grinder to ensure 100 percent penetration.**
- **Don't overweld.** Fat, excessive welds don't gain anything, and they aren't cost effective.
- **Don't weld a continuous bead if the repair piece will be bearing only a light load and the weld doesn't need to be beefy.** Instead, weld an inch-long bead, skip an inch, weld another inch-long bead, and so on. That strategy reduces the risk of distortion in the base metal.
- **Weld in the flat position whenever possible.** Why choose to weld your repair out of position if you can take a few extra minutes and set up the weld so you're in an ideal position?
- **If you have to make multiple passes, alternate from the front to the back of the joint to help prevent distortion and reduce welding stresses on the base metal.**

Getting Ready to Make Repair Welds

You've carefully considered your repair weld and are ready to pull the trigger (no pun intended). In most ways, setting up a repair weld is the same as making any other type of weld: You need to get the welding area ready and make sure you have all the necessary tools and materials on hand. The following sections guide you through those steps.

Preparing your repair piece and work area

The first and most important step in preparing a work area for a repair weld is ensuring you meet all safety standards. Every last one of them! If you're not familiar with welding safety, flip to Chapter 3 right away to get the scoop on all the relevant safety precautions and proper welding safety gear.

Take the time (and apply the elbow grease if necessary) to clean up the metals you'll be welding. Cleaning is an underrated part of any welding repair job. When you're fabricating something new in your welding shop, the metal you're working is more likely to be clean and free of contaminants such as paint, oil, and grease. However, most repair work occurs on metals that have been in use for quite a while and probably need a good cleaning before you can weld them successfully. A few minutes spent with a rag, some solvent, a wire brush, or a grinder can make a huge difference in the quality of your repair weld.

Making the welding area as simple and uncluttered as possible is another important step for success. Is the piece you want to repair tucked in among several other pieces? If so, can you take a little time and simplify the area before you try to make your repair? Sure, you may feel like you've wasted those 30 minutes by moving wiring, hoses, and other pieces of metal away from your repair piece, but it's half an hour well spent if you can avoid damaging those other components.

If the piece you're repairing is in an awkward spot and you can't reach the back of it, consider using a *back-up strip* to ensure good penetration. Ask about them at your welding supply store, and check one out in Figure 18-2.

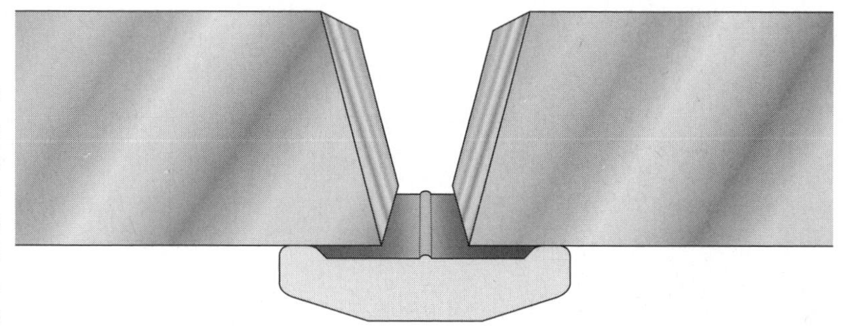

Figure 18-2: An example of a back-up strip with a steel spacer.

Gathering your equipment and tools

Because welding repairs can sometimes be pretty involved — and include some unexpected twists and turns — you want to have all your equipment and tools laid out and close at hand. Here's what I recommend having nearby when you start a repair project.

- **Your welding machine:** An obvious one, I know.
- **Electrodes and filler rods that suit the repair job:** Check out the following section for more on choosing the right electrodes and rods.
- **Grinders:** Grinders always come in handy when you're making repairs because they're good for cleaning up metal and also taking care of jagged, sharp edges. And if you can't get a repaired piece to fit exactly right, a grinder can be a lifesaver.
- **Proper lighting:** Being able to see clearly is more important than ever when you're trying to make a precision weld to repair a broken piece of metal.
- **Tape measure and soapstone:** Measure twice, weld once.

- **Center punch:** These handy little tools are great for getting a cut started in a piece that needs to be cut before it can be modified or repaired.
- **Ventilation or respirator equipment:** Because repair work often involves more contaminants than fabrication does, you want to make sure you're protecting your lungs.
- **Materials for paint prep and painting:** After you've repaired a piece, you may want to get a coat of paint on it soon after it cools off to prevent rust and other corrosive influences.

Selecting filler rods and electrodes

When it comes to welding repair work, all electrodes and filler rods aren't created equal. The following sections give you a rundown of the ones I like best for making welding repairs.

6010 and 6011

These stick welding electrodes are sometimes called *fast fill* or *fast freeze,* meaning they melt easily and harden quickly — great for repair work. With these electrodes, you can expect deep penetration, very little *slag* (the material left over from welding), and an easy-to-control arc. A few other advantages of 6010s and 6011s include

- Great for strong maintenance welds.
- Work well for vertical-up and overhead position welds.
- Useful for galvanized, painted, rusty, or unclean metals. (Just make sure you have the right kind of ventilation or a respirator.)
- Provide great penetration on square butt joints.
- Work on sheet metal.
- Great for welding small-diameter pipe.

When you're selecting current and polarity, remember that 6010 electrodes can be used with DC reverse polarity only. The 6011 electrodes can be used with AC, DC reverse polarity, or DC straight polarity. Head to Chapter 6 for more on types of current.

Low hydrogen electrodes

Low hydrogen electrodes include a few different types of electrodes that produce excellent-quality welds (often called *x-ray-quality* welds). The two most popular are the 7018 and 7028 electrodes. These electrodes can weld thick sections of mild and alloy steel where shrinkage would cause other electrodes to crack. Welds made with these electrodes assume the properties of the metal you're working with and won't crack even on medium to high carbon steel (see Chapter 2). They produce extremely high ductility in the finished repair welds. Both 7018 and 7028 electrodes require that you remove all slag from each weld bead before laying down a new bead. You must always use them with DC reverse polarity. You also have to maintain a short arc, or *porosity* (tiny holes in the weld) can result.

The 7018 electrode works in all positions. Its coating includes iron powder, which becomes part of the weld and actually strengthens it. The 7018 does cause heavy slag, but it's easy to remove.

The 7028 electrodes work really well on deep groove joints. They're considered fast fill (see the preceding section), so they melt quickly. However, one downside is that you can only weld with them in the horizontal and flat positions.

Both the 7018s and 7028s are shipped and sold in sealed containers, so don't allow them to be exposed to moisture before you use them. When these electrodes absorb moisture, they don't weld as well, and you're more likely to end up with porous welds.

Gas welding filler rods

A lot of repair work done with gas welding is done *autogenously* (without filler rods). However, you can find filler rods available for gas welding steel and cast iron. The usual rod length is 36 inches, and the diameters are between 1⁄16 and 5⁄16 inch. The steel rods are copper coated so that they don't rust. The copper coating has no effect on the finished weld. Following are a few gas welding filler rods as they're classified by the American Welding Society (AWS).

- **(RG-60):** These options are great general-purpose gas welding filler rods and can work really well on repairs. You can use them on wrought iron, carbon steel, and low-alloy steel.
- **(RG-65):** These rods are high on the list for repairing sheet metal, pipes, and low-alloy steel. The tensile strength they produce is 65,000 pounds per square inch (not bad!).
- **(RG-45):** This rod is used for mild steel and wrought iron only. It produces welds with only 40,000 pounds per square inch.

Considering Cracks

Cracks are the most serious type of defect that occurs in welding. Cracks create a serious reduction in the strength of a material, and you have to repair them if you want maximum performance out of your metals. A lot of cracks can be seen on the surface of a metal, but some, called *subsurface cracks*, lurk below where you can't spot them. You can check out an example of cracking in Figure 18-3 and read more about various kinds of cracks in Chapter 21.

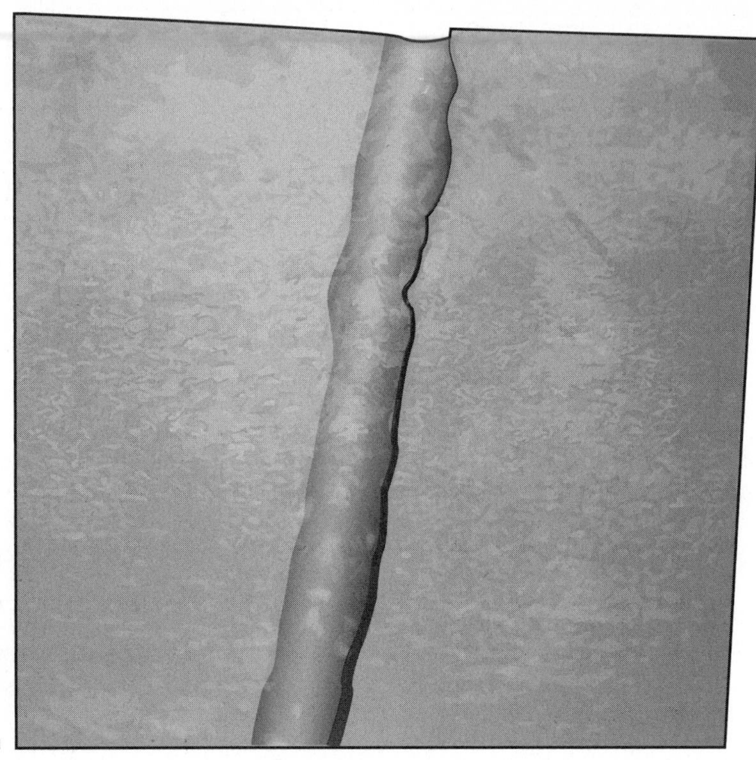

Figure 18-3: Example of a crack along a weld.

Welding can cause cracks in many ways, including the following:

- **Higher carbon content in the base metal than in the filler material.**
- **Too-rapid heating or cooling of the base metal.**
- **Contaminants in the metal or the filler metal.**
- **Dirty electrodes.**
- **Hydrogen incorporated in the weld.** This hydrogen can come from moisture in the air (the most common source), the flux or shielding gas, or the surface of the base metal. If you can control the cooling process after welding by postheating the entire piece and letting it cool off at a uniform rate, you can reverse a hydrogen incorporation problem. Welding in a low-humidity environment can also help.
- **Thick plates (more than ½ inch thick).**
- **Metal-to-metal contact between thick plates prior to welding.** This situation causes all the stress to be put into the weld, which can result in cracking. To reduce this problem, use wire spacers placed between the plates to ensure there's enough of a gap for movement before, during, and after the welding process.

To properly fix cracks and prevent them from growing during the welding process, try drilling small holes at each end of the crack with an electric drill and a drill bit made for use on metal. That keeps the crack from spreading. Another great tactic is to preheat and postheat the base metal before welding so that the temperatures change more slowly; rapid heating and cooling can make cracks much worse.

For details on fixing cracks in cast iron, which is a metal that cracks frequently, check out Chapter 15.

Part VI
The Part of Tens

In this part . . .

This part is always hard to resist for *For Dummies* readers. It features several chapters of handy lists that you can read quickly. That doesn't mean they don't contain some really useful information — quite the contrary — but the info is presented in a fun, easy-to-digest way. My favorite is Chapter 21, which tells you what common welding errors you should be on the lookout for as you weld. (You can't identify and fix a problem if you don't know what it looks like, after all.) But the part also covers signs you're making good welds, plus handy welding tools and the benefits of welding certification. Be sure to check out the glossary, too!

Chapter 19

Ten Tools Every Welder Wants

In This Chapter

- Checking out tools on every welder's wish list
- Storing all your tools in a toolbox

In Chapter 4, you can read all about the tools and other equipment you need to get your welding area set up right. Those are the necessary items. This chapter is all about the tools that welders *want* to have in their shop. They make welding easier, and in many cases, they make the welding process — from start to cleanup — more enjoyable. If you're new to welding and you have a birthday coming up, you're going to want to start dropping hints about these tools to your family.

As with every single tool or piece of equipment I mention in this book, you need to read all the information (instructions and safety details) included with the tool before you use it. You should also read Chapter 3 to brush up on welding safety before trying to use any of this chapter's items. Tools can be extremely dangerous if you don't use them correctly.

You're going to be spending a lot of time with your tools if you really get into welding, so treat them well. Be sure you buy tools that are the right size because tools that are too small or too large for you can be uncomfortable or dangerous. Keep your tools clean, and wipe them down with a rag after each use. If your tool has a blade, keep it sharp. When you're carrying or transporting tools, do so in a way that doesn't result in injury for you or anyone nearby. And finally, if your tools break down, get them repaired properly or replace them. Don't just slap on some duct tape and hope for the best!

4½-Inch Grinder

Grinders are really dynamic tools, and they're a cinch to use after you've had some practice. I love having a 4½-inch grinder in my welding shop. You rarely get all the way through a welding project without having to thoroughly clean or otherwise alter the surface of a metal, and a grinder is an excellent way to get that done in a fast, efficient way. You may be amazed at how often you need to remove rust, grind a groove into something, or smooth down a rough surface.

Part of the magic of a grinder is the variety of attachments you can use with it. Need to smooth something out? Use your flap sanding disc. Want to clean up a piece of cast iron before you weld it? Try your wire brush attachment. Need to grind down some aluminum or another *nonferrous* (iron-free) metal? Attach your nonferrous grinding disc and get started. You can buy a good new grinder for about $100.

Like all power tools, grinders must be properly grounded to prevent electrical shock. Don't use your grinder with excessive force, and if it feels or smells hot, you're working it too hard (so back off). If you use an extension cord, make sure the cord is rated to handle the electrical load necessary to run the tool without damaging the tool or the cord.

Hacksaw

A *hacksaw* is a small, versatile, handheld saw that allows you to use all kinds of different blades designed to cut various materials (including a range of metals). Hacksaws are relatively inexpensive, and you'll probably enjoy having one around. You can pick up a hacksaw for about $10 at your local hardware store or home improvement warehouse, and replacement blades are usually less than $5 for a pack of two or three.

When you use a hacksaw, make sure the teeth point forward so that you're cutting on the stroke away from yourself. This setup helps you prevent inadvertent injury. Plus, starting the cut going away from your body is easier.

Air Compressor

Many of the welding processes, particularly stick welding (see Chapters 5 and 6), can create a mess and leave your projects (and work area) covered with particles. If you have an air compressor with a blowgun attachment, you can use pressurized air to blow that stuff off on the floor so that you can easily broom it up later. That's a lot quicker than having to wipe down or brush off all those surfaces. (Of course, if you have pneumatic tools, you need an air compressor anyway.)

You should be able to pick up a quality air compressor for about $180, and the ones available at your local hardware store are good and strong enough for most new welders.

Air compressors are also very useful to have around if you think you may need to put air in the tires of vehicles, carts, or anything else along those lines.

⅜-Inch Electric Drill

The ⅜-inch electric drill (shown in Figure 19-1) is one of the most common and most popular drills available, and for good reason; you can use this kind of drill for all kinds of things in your home and shop. It's great for drilling small holes in your welding jobs where necessary, and as you begin to take on a wider and wider range of projects, that need pops up more and more often.

Figure 19-1: A standard corded ⅜-inch drill.

There are more than a dozen manufacturers of good-quality ⅜-inch electric drills, so read a few online reviews and ask your welding friends which brands they like before making your purchase. You can find cordless or corded options; I'm partial to the corded models, but if portability is tops on your priority list, a cordless model may be best for you. You can find these drills in a range of prices, from $30 all the way up to about $100.

Wrench Set

When you're assembling or disassembling the parts of a welding project, a good wrench set can certainly come in handy. They're also great for adjusting your tools and other equipment. You can always just stick with an adjustable wrench, which is a very versatile tool, but if you really want a perfect fit, sometimes the adjustable option just doesn't cut it.

I recommend getting a combination wrench set, which includes box- and open-end wrenches. *Box-end* wrenches have a closed loop on the end that you slide down over a bolt before applying force to the wrench. An *open-end wrench,* on the other hand, has more of an open jaw that you can use to slide onto the head of a bolt from the side. Combination wrench sets have a box end on one end and an open end on the other. Both ends can be helpful, so if you've been particularly good this year, when Christmas rolls around you should ask for a set of combination wrenches. (You can see a set of combination wrenches in Figure 19-2.) You (or Santa) can expect to spend about $25 on a set of good wrenches.

Steel Sawhorses

In Chapter 4, I tell you that a welding table is a necessity for any welding shop, and that's the truth. You simply have to have a welding table if you want to get serious about welding. However, in addition to a table, you may also want to get a set of sturdy steel sawhorses like the ones in Figure 19-3. They can provide another means for getting your work up off the floor, making your work on certain objects or projects much easier. And believe me, if you're comfortable when you're welding, you produce much better work (and enjoy doing it)!

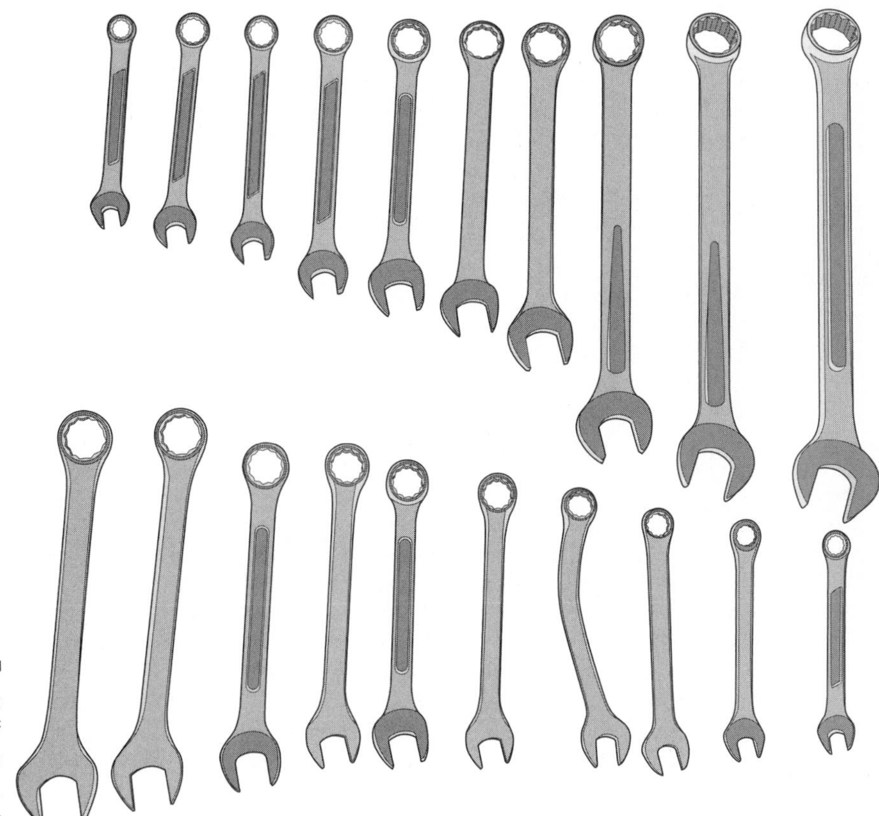

Figure 19-2: A set of combination wrenches.

Figure 19-3: Steel sawhorses.

If necessary, you can also use steel sawhorses as a base for a portable welding table, but don't plan on using them day in and day out for that purpose because it's just not as efficient or as safe. You can get a set of steel sawhorses for as little as $30 at a hardware store or home improvement superstore.

I don't recommend using wooden sawhorses for welding purposes. They're just not nearly as strong or durable as their steel counterparts.

Cutoff Saw

If you haven't worked with a lot of metals in the past, you may not be familiar with cutoff saws, which are a little different from the saws you use to cut other materials such as wood. Instead of using a blade, *cutoff saws* use an abrasive wheel to make cuts. The wheels are like grinding wheels, but they're quite thin; they spin at a very high speed and operate on a dry basis. (Some saws for other types of trades require a wet working area.) You can see a good example of a cutoff saw in Figure 19-4.

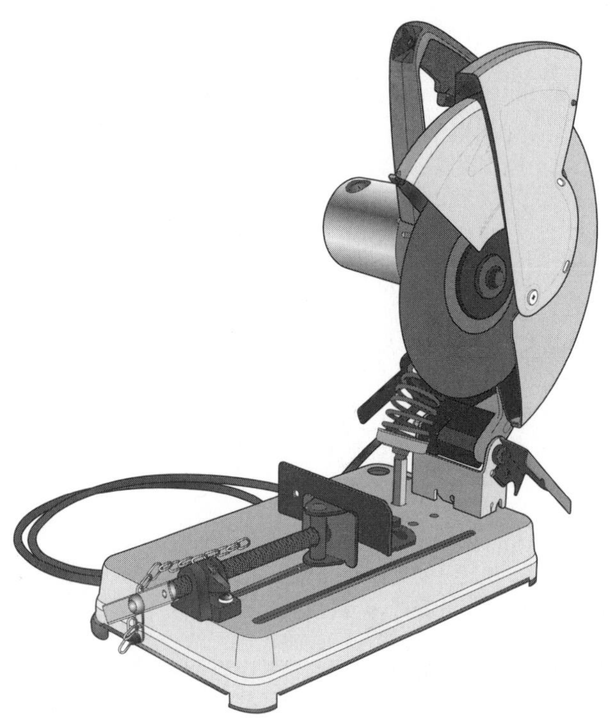

Figure 19-4: A cutoff saw.

I like cutoff saws because the cuts they produce require very little if any cleanup. They're steady and smooth, too, and now that I'm used to having one around, I don't think I would enjoy not having one in my shop. You can pick up a good cutoff saw for around $90.

Before using a cutoff saw, make sure the piece to be cut is clamped very securely into the machine so that it doesn't go flying, and be sure you apply force smoothly at a steady rate while you're cutting. If you apply too much pressure, the cutting blade heats up and can bind to the piece you're cutting.

Bench Grinder

A *bench grinder* is a great piece of equipment for general grinding tasks and for sharpening the tools in your shop. Bench grinders are usually about the size of a microwave, and you can attach them to the top of a workbench (not to the top of your welding table — that takes up too much space). If you haven't seen a bench grinder before, take a look at Figure 19-5. You can buy one for about $50.

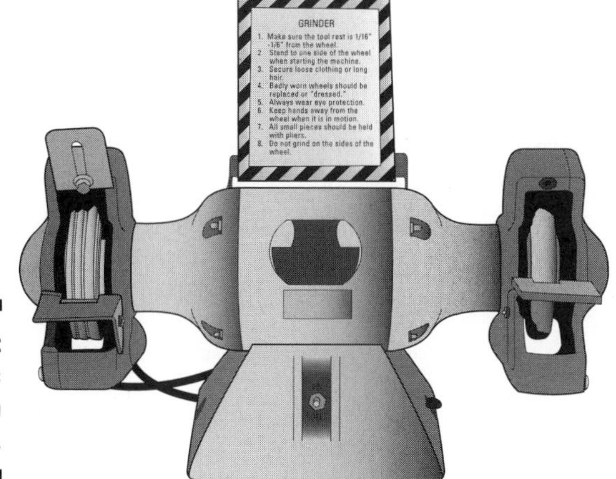

Figure 19-5: A basic bench grinder.

Bench grinders are fitted with replaceable abrasive wheels on both ends. The abrasive wheels are made of a stone material and come in different types for grinding different kinds of metal.

Never grind a nonferrous metal with a ferrous abrasive wheel because it can plug up the metal and explode. Ferrous abrasive wheels are used for grinding cast iron, steel, and stainless steel, and nonferrous abrasive wheels are for aluminum, copper, and brass.

Use a tool rest when you're sharpening tools with your bench grinder. Tool rests give you a good place to stabilize the tool and hold it steady while grinding. Keep the tool rest within 1/16 inch of the abrasive wheel. If for some reason you can't hold the tool you're grinding with both hands, use pliers or vise grips to keep it steady.

I know it may sound counterintuitive, but don't wear gloves when you're using a bench grinder. If the stone happens to catch your glove, it will pull your hand into the grinder and cause serious injury.

Bottle Jack

This option is definitely the kind of tool where the "better to have it and not need it than to need it and not have it" rule applies. If you think you may need to lift heavy materials or pieces in tight spaces, you'll be glad you invested in a $35 bottle jack. A *bottle jack* (shown in Figure 19-6) is a small hydraulic jack that can move a lot of weight in a vertical direction with little effort from the user.

Figure 19-6: A bottle jack.

You can find bottle jacks in different sizes; the bigger ones are generally capable of lifting more weight. The difference in price isn't that much from one jack to another, so I recommend getting a big one.

Toolbox

You really can't beat a durable, trusty toolbox. These heavy-duty boxes store your tools in an organized, safe way. If you don't have a toolbox already, I'd put this item near the top of your wish list. (And if you already have one currently filled with tools for other crafts, such as carpentry or plumbing, consider getting a toolbox exclusively for welding-specific tools).

Toolboxes come in a dizzying array of sizes, from the kind you can fit into the glove box of a car to the kind mounted on wheels that can hold a ton (literally) of tools. Because of the range in sizes, toolboxes can cost anywhere from $10 to $5,000. You can see a medium-sized, basic toolbox in Figure 19-7.

Figure 19-7: A mid-sized metal toolbox.

Most of the bigger boxes have locks on them, or at least some kind of sturdy latch you can use a padlock or combination lock on. I'm a big supporter of locking toolboxes, and because it's not an expensive feature, I recommend you strongly consider it when you go to buy a toolbox.

Chapter 20

(Not Quite) Ten Advantages of Being a Certified Welder

In This Chapter

- Improving job prospects as a certified welder
- Enhancing your skills through certification

Whether you're getting started in welding with intentions to become a full-time welder or are a hobbyist who's thinking about taking his welding efforts to the next level, welding certification is a great option. The main national organization that regulates and administers welding certifications is the American Welding Society (AWS). The AWS has been around for almost 100 years, and according to its Web site, the goal of the organization is "to advance the science, technology and application of welding and related joining disciplines." Most established welders will tell you the same thing about AWS certification: It's an important step for a welder and one that is certainly recognized by the welding industry.

In addition to providing welding certifications, the AWS can also help you adhere to welding codes and standards, and it offers a number of other ways to hone your skills and advance your welding career (if you're interested in getting paid for your welding efforts). Check out its Web site at `www.aws.org`.

If you want to pursue a welding certification from the AWS, get in touch with it — you can find contact information, including e-mail addresses, on its Web site — to find an AWS-sanctioned school. You can then contact the school to talk about how you can work through it to become a certified welder.

Most vocational schools, trade schools, and community colleges include some sort of welding program. In addition to welding certifications, those locations often also offer welding courses (both day and night classes) that can help you to hone your welding knowledge and techniques. The courses usually focus on a relatively narrow range of welding methods, so you can pick and choose the ones you want to get better at or just take them all and become a truly dynamic welder.

The foundation of a welding certification test is a demonstration of skill. You'll be required to show that you can make the kind of weld you're being tested on (in the right position), and that you can do so safely and with the proper setup.

Not everyone passes welding certification tests on the first try. If you don't pass, schedule another test and be sure to talk to the inspector about why you failed.

No matter what you plan to do with your welding skills in the future, I recommend taking a good look at the advantages of getting some sort of welding certification because more often than not, those benefits greatly outweigh the challenges and disadvantages. Read through the following sections and think about what a welding certification can do for you.

More Job Opportunities

Every year the U.S. Department of Labor puts out a report called the Occupational Outlook Handbook (OOH). Among other things, the handbook tells you about the job prospects and earnings information for thousands of different jobs, from nurses to astronomers. Because it's a government report, you can get easy access to it at your local library or online at `www.bls.gov/OCO`.

The OOH contains a section about welders, and it seems like every year the outlook for welders keeps getting better and better. The opportunities and pay just keep going up, and as you can imagine, that isn't true of every field or industry in the handbook. Here's one of many telling statistics in the handbook: In 1998, welders held approximately 367,000 jobs in the United States. In 2008, welders held about 400,000 jobs, and that number is projected to increase between 5 percent and 6 percent again in the next ten years.

By and large, getting a welding job — particularly a welding job with a higher pay range — is much, much easier if you're a certified welder. In coming years, more welding jobs in general and also more jobs with increased earning opportunities will be available, and you'll have much easier access to those positions if you have your certification ducks in a row.

Better Pay

If you want to get paid for your welding work, there's very little argument against getting certified. I have quite a lot of experience in the welding industry, and of the certified and noncertified welders I've known throughout the years, the ones who were working the better-paying jobs were almost always those who had earned a welding certification of some kind.

Every rule has its exceptions, of course, but if you want to make money as a welder, betting heavily on being one of those few exceptions doesn't make sense; after all, if those bets frequently paid off, they'd be the rule, not the exception. And even though getting certified does cost you some money, it's the kind of money you make back quickly after you get the kind of job that goes only to a certified welder.

More Chances for Advancement

Getting one welding certification through the AWS (which I discuss earlier in the chapter) enables you to go back later and get additional, more advanced types of certification. The AWS has a lot of different programs for welders who want to get additional certifications to help them continue to climb up the welding career ladder. Here are a few examples.

- **Welding technician:** A *welding technician* must be able to operate welding equipment correctly, weld successfully using different welding techniques, and pass a written test. These highly skilled welders have received training through a technical school, college, or company-sponsored seminar.
- **Welding supervisor:** *Welding supervisors* manage the work of a staff of welders. If you want to be a welding supervisor, you must have all the basic welding skills along with some technical background on the process you want to supervise.
- **Welding inspector:** This welding position doesn't involve a huge amount of actual welding on a day-to-day basis. A *welding inspector* examines and vets welds with a visual assessment and a hands-on examination involving different inspection tools. As a welding inspector, your job is to make sure a weld or welding project is sound and meets the standards set forth by the AWS.
- **Welding engineer:** A *welding engineer* is the person who makes sure that the metallurgical, mechanical, and structural integrity of any welding job is correct. This certification has some pretty hefty requirements — a welding engineer must pass a four-part examination and have one year of experience as a welder in order to become qualified by the AWS as an engineer.
- **Welding educator:** A *welding educator* must be competent in demonstrating the different welding and cutting skills, and also in explaining the differences among metals.

Certification that Travels with You

According to the U.S. Department of Labor, the areas that offer the highest paying welding jobs are Alaska, Hawaii, Wyoming, and Washington, D.C. Lucky for you, AWS certifications are good in every state in the United States (unlike some of the other skilled trades). If you're a certified welder, you can carry that certification from Alaska out to Hawaii, way over to D.C., and then back west to Wyoming, or anywhere else you want to work.

Ability to Join a National Organization

As a certified welder, you can become a full-fledged member of the AWS, and that membership can offer you some really useful opportunities. I know that some trades have national organizations that don't offer a whole lot, but trust me when I say that the AWS is an organization that's worth joining and being a member of for as long as you want to be a welder. And I'm not alone: The AWS has more than 55,000 members, and that number is growing all the time. It includes welders, welding engineers, welding inspectors, and just about any other welding-related professional you can think of.

The annual dues for an individual membership to the AWS are under $100 (plus a nominal one-time initiation fee), and for that money you get a nice package of benefits, including the following:

- A subscription to a welding publication called *Welding Journal*
- Opportunities to network with other welders and welding professionals at the AWS's section meetings
- Discounts on welding equipment (can't beat that!)
- Access to an AWS-related health insurance program, which is very helpful for freelance welders
- Access to the AWS's open job database — great for finding welding jobs
- Discounts on additional AWS certification programs and educational programs put on by the AWS
- Other benefits like discounts on car rentals, credit cards, and more

I know I kind of sound like a walking advertisement for becoming a member of the AWS, but I really do think it's a good idea for welders and believe it's a good value for the price.

Qualification in Specific Areas of Welding

Because the different areas of welding can be very specific, the AWS provides certifications for specific welding types. Say you really want to focus on tig welding and aren't terribly interested in getting good at mig, stick, or any of the other types of welding. Good news: You can get certified specifically for tig welding (although I'm an advocate for getting familiar with several different welding techniques).

You can even certify for incredibly specific welding types, such as oxyfuel welding on copper nickel pipe by using a grade three silver solder, or tig welding nuclear cooling lines. (Yes, I'm serious — you can look it up.)

Increased Confidence in Your Welding Skills

I know, I know. This one doesn't sound nearly as practical as the other advantages to welding certification that I include in this chapter, but there's definitely something to it.

Imagine you're flying in an airplane. Commercial airplanes have about 6 million parts, and more than a few of those parts are welded together using some of the welding techniques I describe throughout this book. As you're sitting in your seat, hurtling through the air about 30,000 feet above the ground, would you prefer that the welds on your airplane's parts were made by a confident welder or by a welder who has some skills but doesn't have very much confidence? I think most people would choose the confident welder.

When you put in the time and hours of practice to earn a welding certification, it shows that you know your stuff and are good enough to meet some pretty strict standards. That should make you feel good about your skills, which gives you confidence when you take on your next welding job. Even better, that confidence snowballs as you get more and more welding practice under your belt, and before long you're feeling great about the welds you're able to make. That kind of confidence can be a powerful tool, and in addition to making you a better welder, it also helps you to enjoy your work much more than if you were constantly second-guessing your skills and finished products.

Listing in the American Welding Society Database

The AWS maintains a database of all its certified welders, and that's a good list to be on. It allows potential employers to find you and easily confirm your certification on the AWS's Web site. It's a very efficient and helpful clearinghouse for certified welders, and I can't think of any disadvantages to being included in the database.

A Head Start on Additional Types of Welding Certification

After you have experience with and certification in one type of welding technique, you can much more easily move on and get certified in another technique (or several other processes). Some basic welding skills and knowledge apply to all kinds of welding, and after you grasp those concepts, you have a leg up on the material for lots of certifications. As with any type of test or certification process, it only gets easier as you become more and more familiar with the ins and outs.

Chapter 21

(Almost) Ten Welding Defects

In This Chapter

- Identifying common welding errors
- Preventing and fixing problems with your welds

If I could wave my magic filler rod and make it so that all your welds would be strong, clean, and sharp-looking, I'd certainly do it. But the last time I went to the welding supply shop and asked to buy a magic filler rod, they looked at me like I was nuts, so for now you just have to live with the fact that some of your welds will be imperfect. Don't worry about it too much; after all, no welder is perfect, and welding can be a tricky endeavor.

What I *can* do is fill you in on some of the most common weld flaws so that when they show up, you realize that you're dealing with the same kinds of challenges that hundreds of thousands of welders have cursed and spat about since the first guy figured out how to strike an arc. These are the kinds of defects that you're likely to notice only after you've finished a weld (either a single pass or a complete weld, depending on the defect). Most are pretty easy to detect, and — thankfully — relatively easy to adjust for and prevent.

To help prevent welding defects before they happen, be sure your welding materials are clean and in good shape before you start a project. The metal you're planning to weld should be free of any material that may contaminate the weld. Remove any grease, paint, or oil from the metal. You should also put the pieces to be welded in place to make sure they fit together and line up properly.

Incomplete Penetration

Incomplete penetration happens when your filler metal and base metal aren't joined properly, and the result is a gap or a crack of some sort. Check out Figure 21-1 for an example of incomplete penetration.

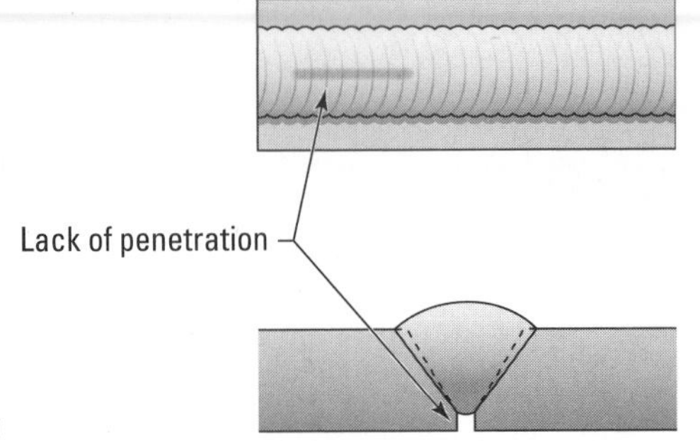

Figure 21-1: A common case of incomplete penetration.

Welds that suffer from incomplete penetration are weak at best, and they'll likely fail if you apply much force to them. (Put simply, welds with incomplete penetration are basically useless.)

Here's a list of the most common causes of incomplete penetration.

- **The groove you're welding is too narrow, and the filler metal doesn't reach the bottom of the joint.**
- **You've left too much space between the pieces you're welding, so they don't melt together on the first pass.**
- **You're welding a joint with a *V*-shaped groove and the angle of the groove is too small (less than 60 to 70 degrees), such that you can't manipulate your electrode at the bottom of the joint to complete the weld.**
- **Your electrode is too large for the metals you're welding.**
- **Your *speed of travel* (how quickly you move the bead) is too fast, so not enough metal is deposited in the joint.**
- **Your welding amperage is too low.** If you don't have enough electricity going to the electrode, the current won't be strong enough to melt the metal properly.

Incomplete Fusion

Incomplete fusion occurs when individual weld beads don't fuse together, or when the weld beads don't fuse properly to the base metal you're welding, such as in Figure 21-2.

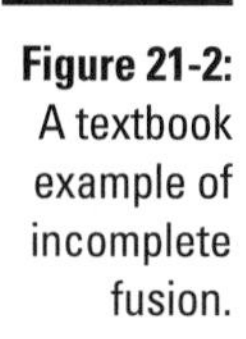

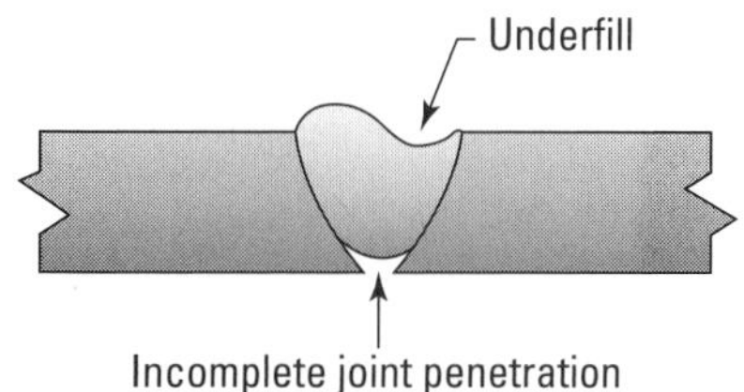

Figure 21-2: A textbook example of incomplete fusion.

The most common type of incomplete fusion is called *overlap* and usually occurs at the *toe* (on the very top or very bottom of the side) of a weld. One of the top causes is an incorrect weld angle, which means you're probably holding the electrode and/or your filler rods at the wrong angle while you're making a weld; if you think that's the case, tweak the angle a little at a time until your overlap problem disappears.

Here are a few more usual suspects when it comes to incomplete fusion causes.

- Your electrode is too small for the thickness of the metal you're welding.
- You're using the wrong electrode for the material that you're welding.
- Your speed of travel is too fast.
- Your arc length is too short.
- Your welding amperage is set too low.

  If you think your incomplete fusion may be because of a low welding amperage, crank up the machine! But be careful: You really need only enough amperage to melt the base metal and ensure a good weld. Anything more is unnecessary and can be dangerous.

- Contaminants or impurities on the surface of the *parent metal* (the metal you're welding) prevent the molten metal (from the filler rod or elsewhere on the parent metal) from fusing.

Undercutting

Undercutting is an extremely common welding defect. It happens when your base metal is burned away at one of the toes of a weld. To see what I mean, look at Figure 21-3.

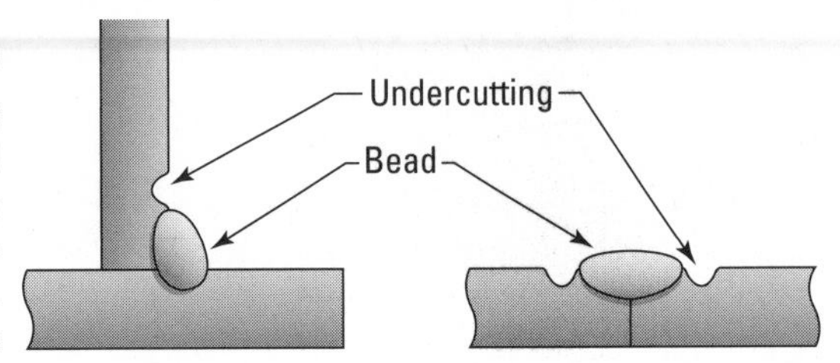

Figure 21-3: A weld suffering from undercutting.

When you weld more than one pass on a joint, undercutting can occur between the passes because the molten weld is already hot and takes less heat to fill, yet you're using the same heat as if it were cold. It's actually a very serious defect that can ruin the quality of a weld, especially when more than 1⁄32 inch is burned away. If you do a pass and notice some undercutting, you must remove it before you make your next pass or you risk trapping *slag* (waste material — see the following section) into the welded joint (which is bad news). The only good thing about undercutting is that it's extremely easy to spot after you know what you're looking for.

Here are a few common causes of undercutting:

- **Your electrode is too large for the base metal you're welding.**
- **Your arc is too long.**
- **You have your amperage set too high.**
- **You're moving your electrode around too much while you're welding.** Weaving your electrode back and forth is okay and even beneficial, but if you do it too much, you're buying a one-way ticket to Undercutting City (which is of course the county seat for Lousy Weld County).

Slag Inclusions

A little bit of slag goes a long way . . . toward ruining an otherwise quality weld. *Slag* is the waste material created when you're welding, and bits of this solid material can become incorporated (accidentally) into your weld, as in Figure 21-4. Bits of flux, rust, and even tungsten can be counted as slag and can cause contamination in your welds.

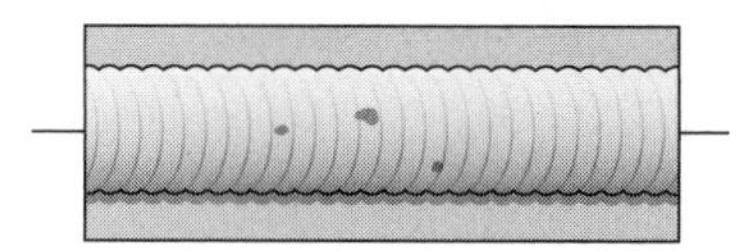

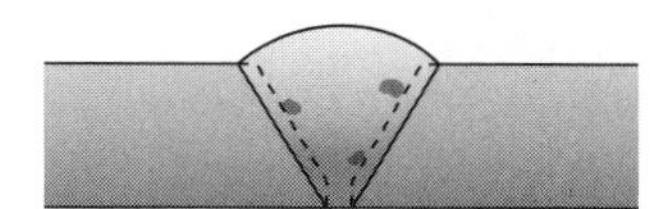

Figure 21-4: A weld with slag inclusions.

Common causes of slag inclusions include

- Flux from the stick welding electrode that comes off and ends up in the weld
- Failure to clean a welding pass before applying the next pass

  Be sure to clean your welds before you go back in and apply a second weld bead.
- Slag running ahead of your weld puddle when you're welding a *V*-shaped groove that's too tight
- Incorrect welding angle
- Welding amperage that's too low

Flux Inclusions

If you're soldering or brazing (also called *braze welding*), flux inclusions can be a real problem. If you use too much flux in an effort to "float out" impurities from your weld, you may very well end up with flux inclusions like those in Figure 21-5. (Head to Chapter 13 for more on brazing and soldering.)

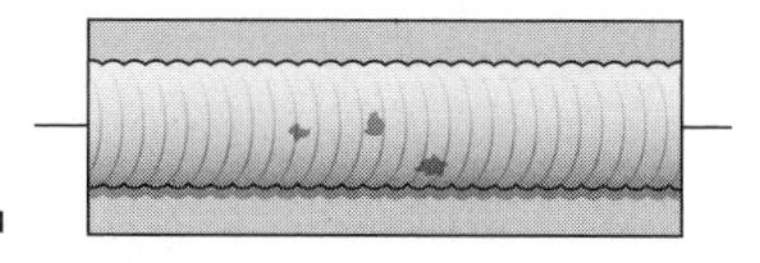

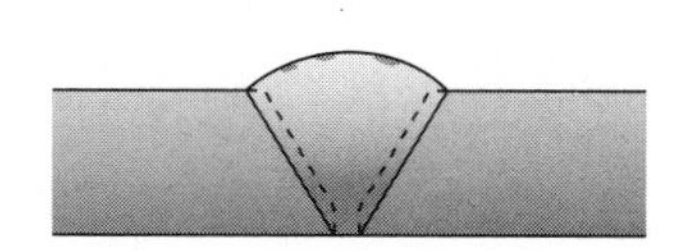

Figure 21-5: Flux inclusions in a finished weld.

If you're working on a multilayer braze weld, flux inclusion can occur when you fail to remove the slag or glass on the surface of the braze before you apply the next layer. When you're soldering, flux inclusion can be a problem if you're not using enough heat. These inclusions are usually closely spaced, and they can cause a soldered joint to leak.

If you want to avoid flux inclusions (and believe me, you do), make sure you do the following:

- **Clean your weld joints properly after each pass.** This task is especially important when you're brazing.
- **Don't go overboard with your use of flux.**
- **Make sure you're using enough heat to melt the filler or flux material.**

Porosity

If you read very much of this book, you quickly figure out that *porosity* (tiny holes in the weld) can be a serious problem in your welds (especially stick or mig welds). Your molten puddle releases gases like hydrogen and carbon dioxide as the puddle cools; if the little pockets of gas don't reach the surface before the metal solidifies, they become incorporated in the weld, and nothing can weaken a weld joint quite like gas pockets. Take a gander at Figure 21-6 for an example of porosity.

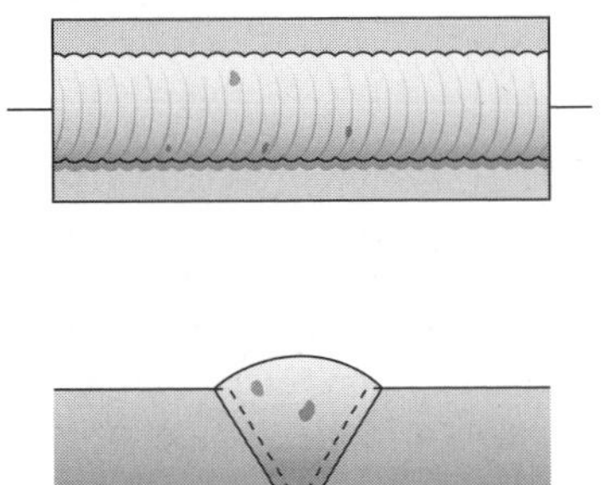

Figure 21-6: A classic case of porosity, which can seriously weaken a weld.

Following are a few simple steps you can take to reduce porosity in your welds:

- Make sure all your materials are clean before you begin welding.
- Work on proper manipulation of your electrode.
- Try using low-hydrogen electrodes.

Cracks

Cracks can occur just about everywhere in a weld: in the weld metal, the plate next to the weld metal, or in any other piece affected by the intense heat of welding. Check out the example of cracking in Figure 21-7.

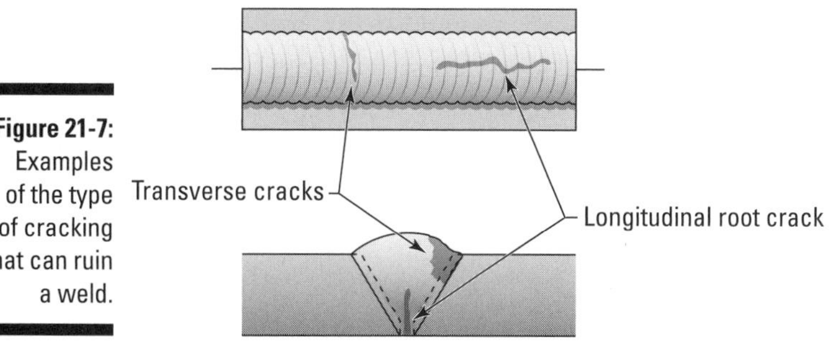

Figure 21-7: Examples of the type of cracking that can ruin a weld.

Here are the three major types of cracks, what causes them, and how you can prevent them.

- **Hot cracks:** This type of crack occurs during welding or shortly after you've deposited a weld, and its cause is simple: The metal gets hot too quickly or cools down too quickly. If you're having problems with hot cracking, try preheating your material. You can also *postheat* your material, which means that you apply a little heat here and there after you've finished welding in an effort to let the metal cool down more gradually.
- **Cold cracks:** This type of crack happens well after a weld is completed and the metal has cooled off. (It can even happen days or weeks after a weld.) It generally happens only in steel, and it's caused by deformities in the structure of the steel. You can guard against cold cracking by increasing the thickness of your first welding pass when starting a new weld. Making sure you're manipulating your electrode properly, as well as pre- and postheating your metal, can also help thwart cold cracking.
- **Crater cracks:** These little devils usually occur at the ending point of a weld, when you've stopped welding before using up the rest of an electrode. The really annoying part about crater cracks is that they can cause other cracks, and the cracking can just kind of snowball from there. You can control the problem by making sure you're using the appropriate amount of amperage and heat for each project, slowing your speed of travel, and pre- and postheating.

Warpage

If you don't properly control the expansion and contraction of the metals you work with, *warpage* (an unwanted distortion in a piece of metal's shape) can be the ugly result. Check out an example in Figure 21-8.

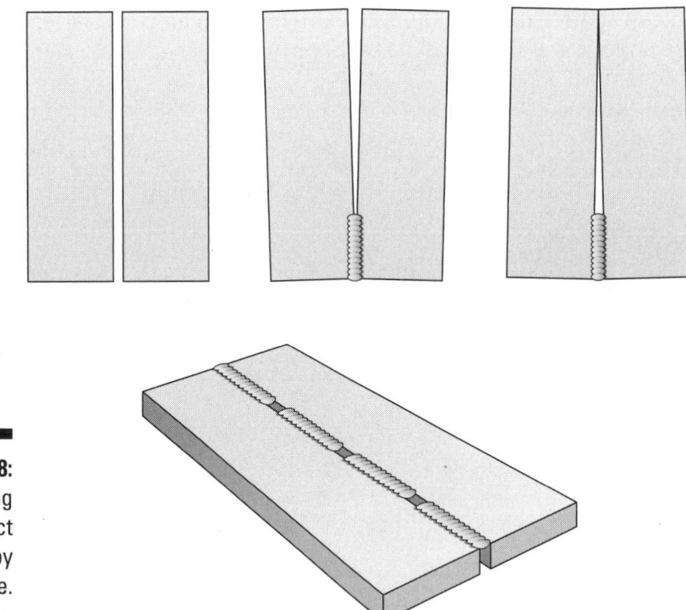

Figure 21-8: A welding project affected by warpage.

If you weld a piece of metal over and over, the chances of it warping are much higher. You can also cause a piece of metal to warp if you clamp the joints too tightly. (If you allow the pieces of metal that make the joint to move a little, there's less stress on them.)

Say you're welding a *T* joint. The vertical part of the *T* sometimes pulls itself toward the weld joint. To account for that movement, simply tilt the vertical part out a little before you weld, so that when it tries to pull toward the weld joint, it pulls itself into a nice 90-degree angle!

The more heat you use, the more likely you are to end up with warpage, so be sure to use only the amount of heat you need. Don't overdo it. Opting for a slower speed of travel while welding can also help to cut down on warpage.

Spatter

Spatter (small particles of metal that attach themselves to the surface of the material you're working on.) is a fact of life with most kinds of welding; no matter how hard you try, you'll never be able to cut it out completely. You can see it in all its glory in Figure 11-5 in Chapter 11.

You can keep spatter to a minimum by spraying with an anti-spatter compound (available at your welding supply store) or by scraping the spatter off the parent metal surface.

Chapter 22

Ten Signs You're Welding Correctly

In This Chapter

- Getting familiar with welding properly
- Checking your work
- Keeping safety in mind

Welding involves a lot of variables: different techniques, different metals, and different positions, to name just a few. But despite all that variety, your goal with any welding project — even with any single weld — is a reliable, safe weld that functions correctly.

To a large degree, welding quality is judged on the intended service of the finished weld. If you're making a weld located in a highly visible area, the quality of your weld depends at least partly on its appearance. Conversely, if you're making a weld to fix a piece of farm machinery, chances are that that weld's quality comes down to whether or not the welded piece is sturdy and secure and allows the machinery to function as it's supposed to. (In that case, it doesn't have to be beautiful to be considered a successful, quality weld.)

No matter what you're shooting for on a welding project or particular weld, a number of signs can indicate that you're doing it right. Read on to discover ten of these useful signs.

The Weld Is Distributed Equally between Parts

If you're welding two pieces of metal together, the molten metal that joins the two pieces should be distributed evenly between them. You control the size of your welds with your *speed of travel* (how fast you make the weld), the type of metal you're depositing, and the thickness of the pieces you're working with. Different projects call for different weld sizes, but despite changes in size, you should always feel good about the project if the weld is distributed equally between the parts you're joining.

Don't over weld! Many people who are new to welding have a tendency to overdo it on their weld sizes. Resist the temptation to make monstrous welds because too much welding can cause the pieces you're working on to become brittle.

The Slag or Shielding Material Doesn't Stick to the Weld

You know you're welding correctly when the waste materials don't stick to your welds. For example, when you're stick welding, you can tell that you have the welding machine set correctly and that the electrodes you're using are dry when the *slag* cools off and starts to peel away easily from the work. You definitely want the residue to come off the welding area nice and easily because you have to remove it before making additional welding passes or prepping the weld to be painted.

When you're mig welding, the residue from your shielding gases should be easy to remove. If it's not, your machine is likely set at the wrong temperature.

Tig welding is the cleanest of all the processes, so you really shouldn't have any waste materials on your tig welds when you're finished. If you do, you started with metals that weren't as clean as they should've been. Be sure all your materials are clean and free of impurities before you start a new tig (or any) welding project.

No Holes or Irregularities on the Weld Surface

Holey weld, Batman! When you look at your welds, the last thing you want to see is *porosity*; a weld with holes is a weak weld, and it tells you that at some point along the way you ran into problems. Holes or other similar irregularities probably mean that you welded a dirty, oily, or wet base metal. It may also mean that you didn't clean off the oxide coating on the base metal before you started welding. Or, if you've been tig or mig welding, porosity can be an indication that you weren't using enough shielding gas. (That's particularly problematic when you weld aluminum.)

The Weld Is Tight

If you're tig welding or oxyfuel welding an *autogenous* weld — that is, the pieces you're welding provide the source of the molten metal with no need for filler material — you can tell you're doing a good job when the weld is nice and tight. A loose weld in that scenario usually means that you're trying to weld two (or more) pieces that don't fit together properly. A tight gap is less important in nonautogenous welds because you can use a filler rod to fill up ill-fitting joints.

When you're making an autogenous weld, your finished product will be much higher quality if the pieces you plan to weld fit tightly together before you start welding. Larger gaps mean weaker finished welds because you don't have enough filler to properly fill the joint.

The Weld Is Leakproof

After you start taking on more and more welding projects, it's only a matter of time before you weld a metal container used to hold a liquid or a gas. I don't know a whole lot of experienced welders who haven't at one point or another made a weld to repair a 55-gallon steel drum, for example.

If you're welding a drum, a tank, or any other kind of container meant to hold a liquid or a gas, a surefire way to recognize a successful weld is a leakproof surface. If the seam in the steel drum that you just welded is leaking water, you definitely didn't get the weld right.

If you weld a container that will eventually be used to hold a pressurized gas, you can always do the soap-bubble test to check the integrity of the weld. After you finish the weld (close all the openings), add air pressure to create pressure inside the container. Then (after the weld cools), simply use a spray bottle to squirt a little soapy water onto the weld. If you see little bubbles in the soapy mixture, that means gas is escaping from the container and you need to try your weld again.

The Weld Has Full Penetration

Penetration is the depth that the molten metal extends from the face of the base metals into the joint. Adequate penetration is completely critical in welding because if molten metal doesn't penetrate deeply enough into the joint, the finished weld simply can't be as strong as it would be if penetration was complete. For example, if you're welding something that's ⅛ inch thick and you get only 1⁄16 inch of penetration, you were able to get the molten metal down only halfway into the joint. That's no good! Over time — and it may not take much time at all — that weld will fail. Any weld that doesn't have the right amount of penetration is a ticking time bomb.

The Weld Has No Undercutting

You can tell that you're doing a good job with your welds if they don't contain any signs of *undercutting,* or a depression on the face of the weld below the surface of the base metal. You may also hear it referred to as *underfill.* This indicator is particularly relevant when you're welding *T* joints, corner joints, or lap joints because if you don't have a full joint, you get a very weak weld.

A number of contributing factors lead to undercutting, and any combination of them can cause the kind of undercutting that really compromises the integrity of your welds. One of the top causes of undercutting is excessive current. You may also get undercutting in your welds if you're not manipulating the puddle correctly, or if you're holding your electrodes at the wrong angle. Inappropriate supplies, such as the wrong filler rods or electrodes, can also cause undercutting.

The Weld Has No Overlap

If for some reason your weld metal isn't melting the edges of the base metal joint and fusing with the base metal, you usually end up with overlap. *Overlap* occurs when the weld metal just piles up in the joint, and welds that suffer from overlap are weak at best.

Like undercutting (see the preceding section), overlap can be caused by a number of factors, including low welding current, improper electrode angle and manipulation, and unusually fast speed of travel. If you start to notice overlap in your welds, make sure your machine current is set correctly, consider the angle at which you're holding your electrode, and remind yourself to take your time when you're welding. These precautions can go a long way toward cutting out your overlap problems.

The Weld Meets Strength Requirements

Although I note earlier in the chapter that different welding projects have different goals that determine what constitutes a quality weld, you pretty rarely find a project where strength isn't at least a minor consideration. If you don't need a finished weld to be strong at all, why not just squirt some white glue in the joint, slap a couple strips of duct tape on it, or maybe just smash a couple pieces of chewing gum in there and hope for the best?

In all seriousness, you need to make sure your welds are strong enough to stand up to whatever forces may be applied to the joint or repaired surface. One of the best ways you can help ensure appropriate strength is to make sure that the filler rod or electrodes you use are rated at a higher strength level than the base metal you're welding. For example, if you're welding on a piece of 40,000-pound steel, use an electrode rated at 60,000 pounds. If you need to step up a little and weld on a piece of 60,000-pound steel, don't continue to use that same electrode. Instead, bump up the strength rating of the electrode to 80,000 pounds and feel confident that you're taking the first step toward ensuring your finished weld will be strong enough to get the job done.

You're Safe and Healthy

What's a surefire sign that you're welding successfully? You're healthy and in one piece when you're finished. You may finish a welding project and end up with the most beautiful, perfectly executed weld in the history of the world. But that weld doesn't mean bubkes if you burned yourself, threw out your back, or damaged your eyes in the process. I would much rather you fail miserably on every single weld you ever attempt than hurt yourself, and you should maintain that same approach every time you turn on your welding machine. Check out Chapter 3 for more on keeping yourself safe while welding.

Chapter 23

Ten Maintenance Tips for Your Welding Equipment and Shop

In This Chapter

- Keeping welding tools and machines in working order
- Tidying up the shop

As the shop owner in charge, you're responsible for maintaining all your tools and equipment in your shop and keeping them in the best possible condition. Doing so helps ensure that your shop runs as efficiently as possible and turns out the best-quality welds it can.

Checking on Your Hand Tools

If you're like most shop owners, you have some time and money invested in hand tools. But how much time have you devoted to taking care of them? The following list shows you how to keep and maintain basic hand tools.

The most important aspect of tool maintenance is keeping the cutting edges sharp and clean.

- **Tape measures:** The most-used tool in your shop is also one of the easiest to clean. All you have to do is occasionally open it up all the way and wipe the blade off with a clean rag to keep it easy to read.
- **Hammers:** Never hit two hammers together because you run the risk of chipping one or both heads. If you notice your hammer has a chip or a piece missing from the striking end, dispose of the hammer; if you strike something with a bad hammer, the chip can come loose and imbed itself in you.

- **Chipping hammers:** Keep the edge of your chipping hammer sharp. If a chip occurs, file or grind it to a smooth, sharp edge.
- **Adjustable wrenches:** Adjustable wrenches need to be clean and have their *worm gears* (the parts that do the adjusting) oiled to keep the wrenches working effectively. If a chip breaks out of the worm gear, get a new wrench; the compromised wrench may adjust on you unexpectedly and cause your grip to slip.
- **Regular wrenches:** Wipe your wrenches clean after completing any job; otherwise, you risk losing the precision performance you need.
- **Screwdrivers:** Use screwdrivers for turning screws only. Using a screwdriver as a chisel may cause the tip to break, which renders the screwdriver useless as a screwdriver and may cause injury if the snapped tip goes airborne. Also, screwdrivers aren't hammers. Trying to hammer something with the handle of a wooden screwdriver may cause the handle to splinter.
- **Pliers:** Occasionally, some metal gets caught in some of the grooved gripping edges of a set of pliers. A quick cleanup with a wire brush helps keep your pliers clear and allows you to grip better.
- **Chisels:** After a lot of hits, the blunt end of the chisel (the part you hit with a hammer) begins to mushroom or spread out. You must remove the mushroom with a grinder to ensure parts of it don't break off and go into the soft part of your hand.
- **Hack saw:** Choosing the correct blade for the job is the first step to any sawing application because using the right blade for the material at hand cuts down on wear and tear. Also, don't saw excessively quickly with a handheld hack saw; it heats up the blade and causes it to lose it sawing ability.
- **Wire brushes:** Make sure your brushes are always clearly labeled with the metal they're intended for; using a brush on the wrong kind of metal shortens the life of the brush and requires more frequent brush replacement.
- **Vise grips:** Like pliers, the grooves in vise grips occasionally need cleaning with a wire brush to keep everything in good working order. You may also need to replace the springs every so often; the heat of welding causes the spring to lose its tension. Replacement springs for your vise grips are available at most welding supply stores.

- **C-clamps:** Make sure the clamping areas of your C-clamps are clean by removing any foreign matter from the threads, or the clamping mechanism can get stuck.
- **Bottle jacks:** Keep the lifting surface of your bottle jack clean and the handle free of any foreign debris. Don't weld close to the lifting shaft of the bottle jack, or your seal may leak.

If you live in a high-humidity area, wipe all your metal tools with a cloth containing a light oil at least once a week, whether you've used them or not, to keep them from rusting. But don't use the oil on any wooden handles.

Taking Care of Power Tools

Maintaining your power tools is a must. I recommend that you keep all of the paperwork that comes with your tools when you purchase them — the manufacturer's technical manual provides all the info you need on proper tool usage and upkeep. Study the directions carefully before attempting any maintenance work on any machine.

The follow suggestions are general for maintenance of your power tools:

- **Lubrication:** Almost all kinds of machines need to be lubricated, especially if they have bearings or gear cases. The manufacturer's instructions will tell you how often to lubricate and what kind of lubricant to use — don't use a lubricant that hasn't been recommended.

  A good way to keep on top of this task is to set up lubrication schedule for all the machinery in your shop. Use a checklist to ensure all your power tools and machinery have been lubricated according to the manufacturer's recommendations.
- **Belt maintenance:** Some older pieces of welding equipment are belt driven, and you need to take particular care to keep the belts on these machines tight. If the belt comes loose, it can run off the pulley and cause serious injury.
- **Motor care:** Make sure you protect the motors of handheld power tools from grinding dust and other debris in the shop. Frequent cleaning of the tools prevents dust accumulation and thus minimizes the chance of it getting into the motor.
- **Tool storage:** Storage cases come with most portable power tools. When you aren't using the tools, they should be in the cases.

You can also help prolong the life of your tools while you're using them by observing a few simple, common-sense precautions:

- Never carry a tool by the electrical cord or air hose.
- Always unplug the tool from the outlet — don't yank the cord from the receptacle.
- Keep the cords clean and away from sharp edges.
- Never leave the cords lying in water, oil, or any other liquid.

Doing Basic Housekeeping in the Shop

In any shop, cleanliness is a very important factor for success. An unclean, disorganized shop isn't safe or productive. That's why you want to make sure your shop is as tidy as possible. The following are some shop considerations to keep in mind:

- Clean out fire hazards on a regular basis. Don't let trash accumulate — dump it every day. Keep greasy and oily rags in a lidded container marked "Dirty Rags" and empty it weekly.
- Wipe up oil spilled on the floor immediately to prevent slipping.
- Make sure all machine guards are in place before operating equipment.
- Don't allow anyone to break off the grounding plugs on your portable electric equipment.
- Before you do any welding, ensure the fire extinguisher is in place and fully charged.

Protecting Your Welding Helmet

Welding helmets are designed to protect you from the ultraviolet and infrared ray of the welding arc — and if you want your helmet to protect you, you should protect it turn. Here are tips to keep up your welding helmet:

- When your helmet is dirty, disassemble, clean, and reassemble it to keep it in the best possible shape.

- Ensure that you can see no white light through any part of your welding helmet when you have it on your head. If you see white light, replace the plastic lens immediately.
- If your helmet becomes damaged, don't tape it together; a taped helmet offers little or no protection. Get a new one.

Seeing to Stick Welding Machine Maintenance

Maintenance of your stick welding machine consists of keeping the machine up and running and replacing parts that wear out during normal operation. You should periodically unplug your welding machine, take off the two side panels if possible, and blow it out with compressed air. Lay the leads out straight and check for any cracks, scuffs, or any imperfections carrying the current to and from your welding operation. Ensure that the spring on the ground clamp is still strong, that the spring and the jaws on your electrode holder are clean and strong, and that the insulation on the electrode itself is in good condition. Clean and wipe the outside surface of your welder after you put it back together.

Stingers or electrode holders regularly wear out and need replacing. The following steps walk you through the replacement process. ***Remember:*** For your protection, read the installation instructions on the box of your new stinger.

1. **Ensure the welding machine is unplugged and loosen the screw that allows the insulator to slide on the cable.**

   You can now you access the connection the between the electrode cable and the stinger.

2. **Loosen the nut holding the electrode cable in the electrode holder, remove the copper wire, and pull the black handle off the cable.**

3. **Remove the handle from the new electrode holder and slide it on the electrode cable, making sure you have the correct end up.**

   You know you have the wrong end up if the handle doesn't fit.

4. **Inspect the copper cable for breakage.**

   If the cable is broken, frayed, or bad, cut it off.

5. **Use a sharp knife to cut the insulation off the wire, leaving a half inch of the new copper exposed.**

   Some new stinger kits come with a piece of copper to wrap around the copper wire before installing in the stringer. Make sure it's tight.

6. **Slide the new handle up the cable and tighten the screw that holds it in place.**

Working on Maintaining Your Mig Welding Machine

Mig wire welding guns are the most handled pieces of equipment, and the ones exposed to the most abuse. They also have the biggest impact on weld quality. The wire welding gun has a nozzle, diffuser, and contact tip located close to the molten pool during welding, which makes them prone to picking up *spatter,* a welding waste material. To remove spatter use needle nose or welding pliers to dislodge it from the tip or the cone.

You should inspect the nozzle several times during any project to ensure the contact tube is tight. I recommend that you check the connections between the wire feeder cable gun and contact tip on a daily basis. If your contact tip isn't the threaded type, you can rotate it occasionally to better conduct electricity.

If you're getting *porosity* (tiny holes) in your welds, check the O-rings between the wire feeder and the gun; they may need replacing, which is very easy and very inexpensive.

The *liner* (which feeds the wire) inside the welding gun component is the hardest thing to check, but when you change your spool of wire, you sometimes get pieces of metal on the wire, and they end up in the liner during changing. Check out Figure 23-1 for a look at a wire welding gun and liner.

Some people blow their liners out with compressed air to get rid of the debris, but you can also just change them. When replacing a liner, lay the old liner out on the floor and then lay the new liner next to it to make sure that the new liner is untwisted; a twisted liner can lead to *birds' nests* (wire coils between the wire feed cable and the drive roll — see Figure 23-2).

You can prevent birds' nests by accurately aligning the inlet guide from the wire feed cable to the rollers. (The liner inlet guide should be as close to the rollers as possible without touching them.)

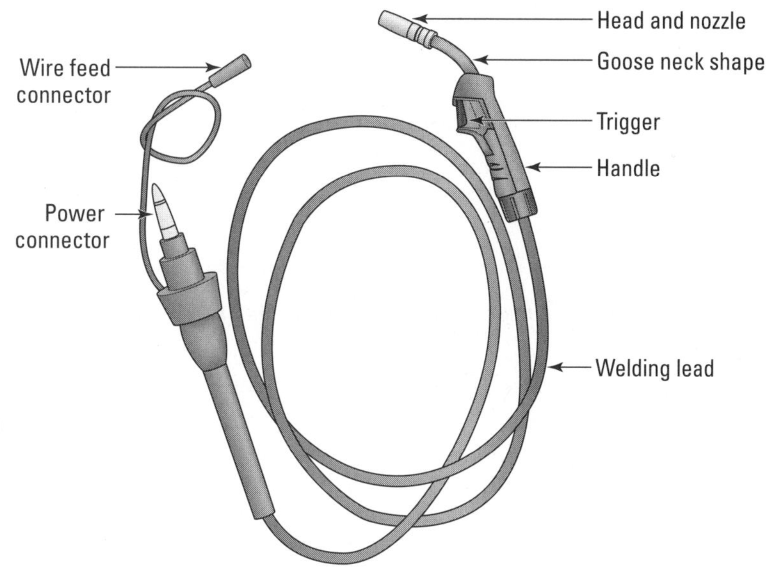

Figure 23-1: Welding gun and liner.

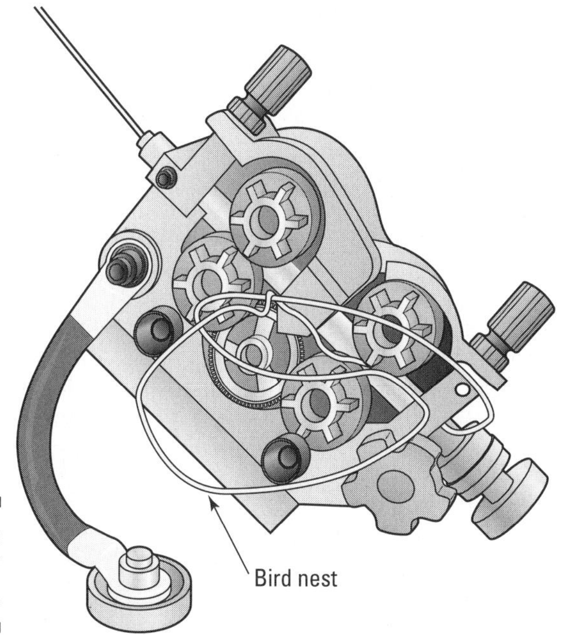

Figure 23-2: A birds' nest.

To fix a bird's nest, just follow these easy steps:

1. **Make sure the machine is unplugged.**
2. **Cut off the wire behind the drive rolls and thread it into one of the holes on the side of the spool.**
3. **Release the pressure on the drive rolls and unscrew the fastener that holds your wire welding gun assembly tight against the rollers.**
4. **Remove the wire welding gun from the machine.**
5. **Disconnect the electrical wire that controls the trigger switch on the gun.**
6. **Laying the assembly flat on the floor, take off the cone, and cut off any wire extending past the contact tip.**
7. **Go to the machine end of your wire welding gun and pull out the wire.**
8. **Unhook the wire from your spool of wire.**
9. **Cut off the end and thread it into the inlet guide through the roller so that it sticks out four to six inches past the machine; tighten up your roller pressure assembly so the wire doesn't move.**
10. **Pick up the machine end of your wire welding gun and thread it on to the wire sticking past the machine; pick up the gun and feed the four to six inches of wire into the liner.**
11. **Reassemble the machine end of the wire welding gun into the machine and tighten the screw holding it in place.**
12. **Hook the trigger wires back up and remove the contact tip.**
13. **Plug in and turn on the welder and pull the trigger until the wire comes out two to three inches past the cone; reinstall the contact tip and cone.**

Tuning Up Your Tig Welding Machine

To keep your tig welding machine in good shape, periodically lay out all the leads on the floor and inspect the ground clamp, the tig torch cable, and the hoses for cracks, abrasions, or any other defects. If you don't find any, coil them up neatly. Inspect the hose from the shielding gas bottle to the back of the machine.

Make sure the tig torch has no cracks or imperfections, and check the ceramic cup. If you have a water cooling tig torch, make sure the water doesn't leak and that the pressure isn't higher than what the manufacturer recommends.

Tig welding is a very clean process, so you don't normally have to blow this machine out.

Taking Care of Your Oxyfuel Equipment

Most oxyfuel equipment maintenance is common sense. The only piece that wears out on an oxyacetylene torch is the tip, and it doesn't really wear out; it just gets bigger from too many cleanings. Here are some general oxyfuel upkeep tips:

- **Make sure you keep all your bottles secured and in the upright position.**
- **After assembling your torch and turning it on, check for leaks by doing a soap test (see Chapter 3).**
- **Never change tips while they're hot.** Trying to change tips while they're hot can result in burns.
- **Keep the orifice at the end of the tip clean at all times.** Clean the tip with tip cleaners, not by scratching the tip on the fire brick or the piece you are welding — you risk less damage this way.
- **Don't use the tip as a hammer.**
- **Always make sure you have flashback arrestors in line on your hoses.** A *flashback arrestor* is a safety mechanism that stops the flow of gas if a flame starts to burn gas in your hoses or other parts of your gas supply system.
- **Inspect the oxyacetylene hoses frequently to ensure they're leak-proof.** If your hose leaks and is unfit for service, replace it. Never use tape or hose clamps to try to repair it, because these repairs just don't do the job.
- **Never use oil on oxyacetylene equipment.** When oxygen comes in contact with oil, it can explode.
- **Keep an eye out for a *creeper gauge,* which is the term for a gauge that magically changes settings without you actually adjusting it.** If you find you have a creeper, send it in to be fixed.

Keeping Your Air Compressor Working

As I note in Chapter 19, an air compressor is a handy tool to have in your welding shop. But like all equipment, you get the best use out of your compressor if you put a little time into maintaining it. Check out these maintenance tips:

- **Clean the intake vents.** Be sure to keep the vent that allows air into your air compressor as clean as possible. Check them daily if you're working in a dirty area.
- **Tighten all fasteners.** If an engine is running, it's vibrating, and vibration loosens nuts and bolts. Be sure to check them regularly and tighten them as necessary.
- **Check hoses.** Make sure your hoses aren't cracked, scuffed, or worn because damaged hoses can soon begin to leak and make your compressor work harder. Don't try to repair them — just replace them.
- **Monitor the air filters.** Check and/or change the air filters regularly if you notice a buildup of dirt or other material on the surface.
- **Change the compressor oil.** Most manufacturers recommend that you change the oil in your compressor every 750 hours. Check your manual to see how often you should change the oil in your air compressor, and make sure you follow that guideline.

Drill Press Maintenance

Earlier in the chapter, I cover the importance of keeping power equipment lubricated, and this consideration is especially important with drill presses such as the one in Figure 23-3. Proper lubrication greatly prolongs the life of your drill press. Lubricate any place that metal touches metal per manufacturer's instructions. The unpainted exterior surface should always have a light film of oil. If your press has a gear rack for adjusting the height of the drill press motor, the teeth should have a light deposit of grease.

Only use lubricants recommend by the manufacturer.

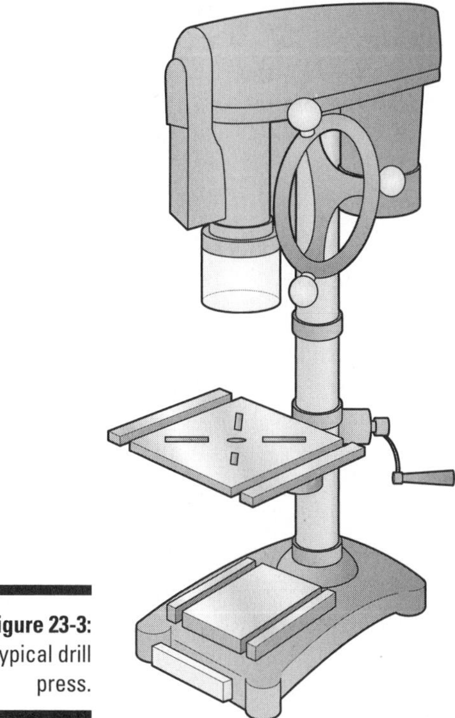

Figure 23-3: Typical drill press.

Glossary

acetone: Flammable liquid commonly used in acetylene cylinders to stabilize acetylene.

acetylene: Combustible gas composed of carbon and hydrogen; used as fuel in the oxyacetylene welding process.

air-acetylene: Flare produced by burning acetylene with air instead of oxygen.

air-arc cutting: Arc cutting process in which the metals to be cut are melted by the heat of the carbon arc.

alloy: Substance composed of two or more elements, of which at least one is a metal.

alternating current (AC): Electric current that reverses its direction at regularly recurring intervals.

ammeter: Instrument for measuring electrical current in amperes.

annealing: Opposite of hardening, done to remove hardness in certain metals where drilling or other machining is desired.

arc blow: The deflection of an electric arc from its normal path due to magnetic forces.

arc brazing: Brazing process wherein heat is obtained from an electric arc formed between the base metal and an electrode, or between two electrodes.

arc cutting: Cutting process in which metal is cut by melting with the heat of an arc between the electrode and the base metal.

arc length: The distance between the electrode tip and the weld puddle.

arc strike: Unintentional arc start outside of the weld bead.

arc welding: Welding process in which fusion is obtained by heating with an electric arc or arcs, with or without the use of filler metal.

austenite: Non-magnetic form of iron characterized by a face-centered cubic lattice crystal structure. It's produced by heating steel above the upper critical temperature and has a high solid solubility for carbon and alloying elements.

axis of a weld: Line through the length of a weld, perpendicular to a cross section at its center of gravity.

backfire: Momentary burning back of a flame into the tip, followed by a snap or pop, and then the immediate reappearance or burning out of the flame.

back pass: Pass made to deposit a back weld.

back up: In flash and upset welding, a locator used to transmit all or part of the upsetting force to the work pieces.

back weld: Weld deposited at the back of a single groove weld.

backhand welding: Welding technique in which the flame is directed toward the completed weld.

backing ring: Metal ring placed inside the seam of a pipe being butt-welded. The ring allows for full weld penetration and maximum weld strength.

backing strip: Piece of material used to retain molten metal at the root of the weld, or increase the thermal capacity of the joint to prevent excessive warping of the base metal.

backing weld: Weld bead applied to the root of a groove joint to ensure complete root penetration.

backstep: Sequence in which weld bead increments are deposited in a direction opposite to the direction of progress.

bare electrode: Consumable, bare electrode used in arc welding with no flux coating.

base metal arc welding: Arc welding process in which fusion is obtained by heating with an unshielded arc between a bare or lightly-coated electrode and the work. Pressure isn't used and filler metal is obtained from the electrode.

base metal: Metal to be welded or cut.

bead: Seam between work pieces that have been joined with welding.

bead weld: Weld composed of one or more string or weave beads deposited on an unbroken surface.

bevel angle: Angle formed between the prepared edge of the metal and a plane perpendicular to the surface of the metal.

billet: Solid bar of metal usually made by forging instead of casting.

block brazing: Brazing process in which bonding is produced by the heat obtained from heated blocks applied to the parts to be joined and by a filler metal having a melting point above 800 degrees F (427 degrees C), but below that of the base metal.

bond: Junction of the welding metal and the base metal.

boxing: Continuing a fillet weld around a corner of a member as an extension of the principal weld.

brazing: Welding process in which a groove, fillet, lap, or flange joint is bonded by using a nonferrous filler metal with a melting point above 800 degrees F (427 degrees C), but below that of the base metal.

braze welding: Welding by using a filler metal that liquefies above 450 degrees C (842 degrees F) and below the solid state of the base metal.

bridging: Defect caused by poor penetration. A void at the root of the weld is spanned by weld metal.

buckling: Distortion caused by the heat of a welding process.

build-up sequence: Order in which the weld beads of a multipass weld are deposited with respect to the cross section of a joint.

butt joint: Joint between two work pieces in which the weld joining the parts is between the surface planes of both of the pieces joined.

butt weld: Weld in a butt joint.

butter weld: Weld consisting of one or more string or weave beads laid down on an unbroken surface to obtain desired properties or dimensions.

capillary action: Action in which adhesion between the molten filler metal and the base metals, along with surface tension of the molten filler metal, causes distribution of the filler metal between the properly fitted surfaces of the joint to be brazed.

carbide precipitation: Unstable condition occurring in austenitic stainless steel which contains carbon in a supersaturated solid solution. Agitation of the steel during welding causes the excess carbon in solution to precipitate.

carburizing flame: Oxyacetylene flame in which there is an excess of acetylene.

cascade sequence: When subsequent beads are stopped short of a previous bead, giving a cascade effect.

case hardening: Surface hardening involving a change in the composition of the outer layer of an iron base alloy by inward diffusion from a gas or liquid, followed by appropriate thermal treatment. Typical hardening processes are carbonizing, cyaniding, carbonitriding, and nitriding.

chain intermittent fillet welds: Two lines of intermittent fillet welds in a T or lap joint in which the welds in one line are approximately opposite those in the other line.

chamfering: Preparation of a welding contour other than that for a square groove weld, on the edge of a joint member.

coalescence: Uniting or fusing of metals upon heating.

coated electrode: Electrode with a flux applied externally by dipping, spraying, painting, or other similar methods. Upon burning, the coat produces a gas which envelopes the arc.

cold weld: Poor penetration of the weld bead, usually less than 5 percent of the bead thickness.

composite electrode: Filler metal electrode used in arc welding, consisting of more than one metal component combined mechanically. It may or may not include materials that improve the properties of the weld, or stabilize the arc.

composite joint: Joint in which both thermal and mechanical processes are used to unite the base metal parts.

concavity: Maximum perpendicular distance from the face of a concave weld to a line joining the toes.

concurrent heating: Supplemental heat applied to a structure during the course of welding.

cone: Conical part of a gas flame next to the orifice of the tip.

consumable electrode: Electrode that also serves as the filler material.

consumable insert: Preplaced filler metal that is completely fused into the root of the joint and becomes part of the weld.

convexity: The maximum perpendicular distance from the face of a convex fillet weld to a line joining the toes.

corner joint: Joint between two metal pieces located approximately at right angles to each other in the form of an L.

corrosion: Gradual chemical attack on metal by moisture, the atmosphere, or other agents.

cover glass: Clear glass used in goggles, hand shields, and helmets to protect the filter glass from spattering material.

covered electrode: Metal electrode with a covering material that stabilizes the arc and improves the properties of the welding metal. The material may be paper, asbestos, and other materials or a flux covering.

crack: Fracture characterized by a sharp tip and high ratio of length and width to opening displacement.

crater: Depression at the termination of an arc weld.

critical temperature: Transition temperature of a substance from one crystalline form to another.

current density: Amperes per square inch of the electrode cross sectional area.

cutting tip: Gas torch tip especially adapted for cutting.

cutting torch: Device used in gas cutting for controlling the gases used for preheating and the oxygen used for cutting the metal.

cylinder: Portable cylindrical container used to store compressed gas.

defect: Discontinuity or discontinuities which, by nature or accumulated effect (for example, total crack length), render a part or product unable to meet the minimum applicable acceptance standards or specifications. This term designates rejectability.

deposited material: Filler metal that has been added during a welding operation.

deposition efficiency: Ratio of the weight of deposited metal to the net weight of electrodes consumed, exclusive of stubs.

depth of fusion: Depth that fused filler material extends into the base metal.

die: Device used in forge welding primarily to form the work while hot and apply the necessary pressure.

die welding: Forge welding process in which fusion is produced by heating in a furnace and by applying pressure by means of dies.

dip brazing: Brazing process in which bonding is produced by heating in a molten chemical or metal bath and by using a nonferrous filler metal having a melting point above 800 degrees F (427 degrees C), but below that of the base metals. The filler metal is distributed in the joint by capillary attraction. When a metal bath is used, the bath provides the filler metal.

direct current electrode negative (DCEN): Arrangement of direct current arc welding leads in which the work is the positive pole and the electrode is the negative pole of the welding arc.

direct current electrode positive (DCEP): Arrangement of direct current arc welding leads in which the work is the negative pole and the electrode is the positive pole of the welding arc.

discontinuity: Interruption of the typical structure of a weldment, such as lack of homogeneity in the mechanical, metallurgical, or physical characteristics of the material or weldment. A discontinuity is not necessarily a defect.

drag: Horizontal distance between the point of entrance and the point of exit of a cutting oxygen stream.

drop through: Filler material that sags through the underside of the weld, caused by either too much heat or poor joint fit.

ductility: Metal property that allows it to be permanently deformed, in tension, before final rupture. Ductility is commonly evaluated by tensile testing in which the amount of elongation and the reduction of area of the broken specimen, as compared to the original test specimen, are measured and calculated.

duty cycle: Percentage of time during an arbitrary test period, usually 10 minutes, during which a power supply can be operated at its rated output without overloading.

dye-penetrating test: Process for checking welds for cracks and other defects. The process consists of three chemicals in solution: a cleaner, a spray-on penetrating red dye, and a white developer solution. After the area is cleaned and the red dye allowed to soak in a few minutes, the developer is sprayed on. Defects show red and smooth areas appear white. After inspection, the cleaning solution can be used to remove the dye.

edge joint: Joint between the edges of two or more parallel or nearly parallel members.

edge preparation: Contour prepared on the edge of a joint member for welding.

effective length of weld: Length of weld throughout which the correctly proportioned cross section exits.

electrode: Wire or rod, bare or covered, through which current is conducted between the electrode holder and the arc.

electrode holder: Device used for mechanically holding the electrode and conducting current to it.

electrode skid: Sliding an electrode along the surface of the work during spot, seam, or projection welding.

embossment: Rise or protrusion from the surface of a metal.

etching: Process of preparing metallic specimens and welds for macrographic or micrographic examination.

face reinforcement: Reinforcement of weld at the side of the joint from which welding was done.

face of a weld: Exposed surface of a weld, made by an arc or gas welding process, on the side from which welding was done.

faying surface: Surface of a piece that is in contact with another piece to which it is joined.

ferrite: Virtually pure form of iron existing below the lower critical temperature and characterized by a body-centered cubic lattice crystal structure. It's magnetic and has very slight solid solubility for carbon.

field weld: Weld done at the site or in the field rather than in a welding shop.

filler metal: Metal to be added in making a weld.

fillet weld: Weld of approximately triangular cross section, as used in a lap joint, joining two surfaces at approximately right angles to each other.

filter gas: Colored glass used in goggles, helmets, and shields to exclude harmful light rays.

flame hardening: Method for hardening a steel surface by heating with a gas flame followed by a rapid quench.

flame softening: Method for softening steel by heating with a gas flame followed by slow cooling.

flash: Metal and oxide expelled from a joint made by a resistance welding process.

flash burn: Burn caused by ultraviolet light radiation from the arc in arc welding. More severe than a sunburn.

flash welding: Resistance welding process in which fusion is produced, simultaneously over the entire area of abutting surfaces, by the heat obtained from resistance to the flow of current between two surfaces and by the application of pressure after heating is substantially completed. Flashing is accompanied by expulsion of metal from the joint.

flashback: Burning gases within the torch or beyond the torch in the hose, usually with a shrill, hissing sound.

flat position: Position in which welding is performed from the upper side of the joint and the face of the weld is approximately horizontal.

flow welding: Process in which fusion is produced by heating with molten filler metal poured over the surfaces to be welded until the welding temperature is attained and the required filler metal has been added.

flux: Cleaning agent used to dissolve oxides, release trapped gases and slag, and to cleanse metals for welding, soldering, and brazing.

flux cored wire: Electrode that contains flux within a wire tube.

flux cored arc welding (FCAW): Welding process that uses a tubular weld wire with an internal flux with or without an externally supplied gas shielding.

forehand welding: Gas welding technique in which the flare is directed against the base metal ahead of the completed weld.

forge welding: Welding process in which fusion is produced by heating in a forge or furnace and applying pressure or blows.

full fillet weld: Fillet weld in which the size is equal to the thickness of the thinner member joined.

furnace brazing: Process in which bonding is produced by the furnace heat and a nonferrous filler metal having a melting point above 800 degrees F (427 degrees C), but below that of the base metals.

fusion: Thorough and complete mixing between the two edges of the base metal to be joined or between the base metal and the filler metal added during welding.

fusion zone (filler penetration): Area of base metal melted as determined on the cross section of a weld.

gas metal arc (mig) welding (GMAW): Arc welding process using a wire electrode that is also the filler material. An inert gas is distributed over the weld areas to shield the molten metal from oxygen.

gas pocket: Weld cavity caused by the trapping of gases released by the metal when cooling.

gas tungsten-arc (tig) welding (GTAW): Arc welding process in which fusion is produced by heating with an electric arc between a tungsten electrode and the work while an inert gas forms around the weld area to prevent oxidation. No flux is used.

gas welding: Process in which the welding heat is obtained from a gas flame.

globular transfer (arc welding): Type of metal transfer in which molten filler metal is transferred across the arc in large droplets.

goggles: Device with colored lenses which protect the eyes from harmful radiation during welding and cutting operations.

groove: Opening provided between two members to be joined by a groove weld.

groove angle: Total included angle of the groove between parts to be joined by a groove weld.

groove face: Surface of a member included in the groove.

groove radius: Radius of a J or U groove.

groove weld: Weld made by depositing filler metal in a groove between two members to be joined.

ground connection: Connection of the work lead to the work.

guided bend test: Bending test in which the test specimen is bent to a definite shape by means of a jig.

hand shield: Device used in arc welding to protect the face and neck. It is equipped with a filter glass lens and is designed to be held by hand.

hard facing: Form of surfacing in which a coating or cladding is applied to a surface for the main purpose of reducing wear or loss of material by abrasion, impact, erosion, galling, and cavitations.

hard surfacing: Application of a hard, wear-resistant alloy to the surface of a softer metal.

hardening: Increasing the hardness of metal by suitable treatment, usually involving heating and cooling.

Heliarc: Trademark of Linde; gas tungsten arc welding.

heat affected zone: Portion of the base metal whose structure or properties have been changed by the heat of welding or cutting.

heat sink: Mass of metal, water-soaked rag, or other heat-absorbing material placed so it absorbs heat, thus preventing overheating of a component or area. The use of a heat sink can prevent or limit burnthrough or warpage.

heat time: Duration of each current impulse in pulse welding.

heating gate: Opening in a thermit mold through which the parts to be welded are preheated.

heat treating: Process that adds strength and brittleness to metal, involving controlled heating and cooling of the metal to achieve the desired change in crystalline structure. Almost all metals have a critical temperature that changes their grain structure, but not all metals can be heat-treated.

home time: Time in which pressure is maintained at the electrodes after the welding current has stopped.

horizontal weld: Bead or butt welding process with its linear direction horizontal or inclined at an angle less than 45 degrees to the horizontal, and the parts welded being vertically or approximately vertically disposed.

horn: electrode holding arm of a resistance spot welding machine.

hot short: Condition that occurs when a metal is heated to that point, prior to melting, where all strength is lost but the shape is still maintained.

hygroscopic: Readily absorbing and retaining moisture.

impact test: Test in which one or more blows are suddenly applied to a specimen. The results are usually expressed in terms of energy absorbed or number of blows of a given intensity required to break the specimen.

impregnated tape metal arc welding: Arc welding process in which fusion is produced by heating with an electric arc between a metal electrode and the work. Shielding is obtained from decomposition of impregnated tape wrapped around the electrode as it is fed to the arc. Pressure is not used, and filler metal is obtained from the electrode.

induction brazing: Process in which bonding is produced by the heat obtained from the resistance of the work to the flow of induced electric current and by using a nonferrous filler metal having a melting point above 800 degrees F (427 degrees C), but below that of the base metals.

induction welding: Process in which fusion is produced by heat obtained from resistance of the work to the flow of induced electric current, with or without the application of pressure.

inert gas: Gas that does not normally combine chemically with the base metal or filler metal.

interpass temperature: In a multipass weld, the lowest temperature of the deposited weld meal before the next pass is started.

joint: The portion of a structure in which separate base metal parts are joined.

joint penetration: Maximum depth a groove weld extends from its face into a joint, exclusive of reinforcement.

kerf: Space from which metal has been removed by a cutting process.

keyholing: Usually occurs while burr-welding two very thin pieces of aluminum. A small hole melts all the way through but is filled with filler rod.

lap joint: Joint between two overlapping members.

layer: Stratum of weld metal, consisting of one or more weld beads.

liquidus: Lowest temperature at which a metal or an alloy is completely liquid.

local preheating: Preheating a specific portion of a structure.

manifold: Multiple header for connecting several cylinders to one or more torch supply lines.

martensite: Microconstituent or structure in quenched steel characterized by an acicular or needle-like pattern on the surface of polish. It has the maximum hardness of any of the structures resulting from the decomposition products of austenite.

mash seam welding: Seam weld made in a lap joint in which the thickness at the lap is reduced to approximately the thickness of one of the lapped joints by applying pressure while the metal is in a plastic state.

melting point: Temperature at which a metal begins to liquefy.

melting range: Temperature range between solidus and liquidus.

melting rate: Weight or length of electrode melted in a unit of time.

metallizing: Method of overlay or metal bonding to repair worn parts.

mig welding: Arc welding process using a wire electrode that is also the filler material. An inert gas is distributed over the weld areas to shield the molten metal from oxygen.

mixing chamber: Part of a welding or cutting torch in which the gases are mixed for combustion.

multi-impulse welding: Making spot, projection, and upset welds by more than one impulse of current. When alternating current is used, each impulse may consist of a fraction of a cycle or a number of cycles.

neutral flame: Gas flame in which the oxygen and acetylene volumes are balanced and both gases are completely burned.

nick break test: Method for testing weld soundness by nicking each end of the weld, then giving the test specimen a sharp hammer blow to break the weld from nick to nick. Visual inspection will show any weld defects.

nitriding: Surface-hardening process for certain steels, in which nitrogen is introduced in contact with anhydrous ammonia gas in the 935-to-1000 degrees F (502-to-538 degrees C) range. Quenching is not required.

nonferrous: Metals that contain no iron. Aluminum, brass, bronze, copper, lead, nickel, and titanium are nonferrous.

normalizing: Heating iron-base alloys to approximately 100 degrees F (38 degrees C) above the critical temperature range, followed by cooling to below that range in still air at ordinary temperature.

nugget: Fused metal zone of a resistance weld.

open circuit voltage: Voltage between the terminals of the welding source when no current is flowing in the welding circuit.

overhead welding: Position in which welding is performed from the underside of a joint and the face of the weld is approximately horizontal.

overlap: Protrusion of weld metal beyond the bond at the toe of the weld.

oxidizing flame: Oxyacetylene flame in which there is an excess of oxygen. The unburned excess tends to oxidize the weld metal.

oxyacetylene cutting: Oxygen cutting process in which the necessary cutting temperature is maintained by flames obtained from the combustion of acetylene with oxygen.

oxyacetylene welding: Welding process in which the required temperature is attained by flames obtained from the combustion of acetylene with oxygen.

oxy-arc cutting: Oxygen cutting process in which the necessary cutting temperature is maintained by means of an arc between an electrode and the base metal.

oxy-city gas cutting: Oxygen cutting process in which the necessary cutting temperature is maintained by flames obtained from the combustion of city gas with oxygen.

oxygen cutting: Cutting ferrous metals by means of the chemical action of oxygen on elements in the base metal at elevated temperatures.

oxygen gouging: Application of oxygen cutting in which a chamfer or groove is formed.

oxy-hydrogen cutting: Oxygen cutting process in which the necessary cutting temperature is maintained by flames obtained from the combustion of city gas with oxygen.

oxy-hydrogen welding: Gas welding process in which the required welding temperature is attained by flames obtained from the combustion of hydrogen with oxygen.

oxy-natural gas cutting: Oxygen cutting process in which the necessary cutting temperature is maintained by flames obtained by the combustion of natural gas with oxygen.

oxy-propane cutting: Oxygen cutting process in which the necessary cutting temperature is maintained by flames obtained from the combustion of propane with oxygen.

pass: Weld metal deposited in one general progression along the axis of the weld.

peening: Working metals by means of hammer blows. Peening tends to stretch the surface of the cold metal, thereby relieving contraction stresses.

percussive welding: Resistance welding process in which a discharge of electrical energy and the application of high pressure occurs simultaneously, or with the electrical discharge occurring slightly before the application of pressure.

perlite: Lamellar aggregate of ferrite and iron carbide resulting from the direct transformation of austenite at the lower critical point.

pickling: Putting metal in a diluted acid or other chemical to clean unwanted matter from its surface.

pitch: Center to center spacing of welds.

plating: Outer coating of chromium, copper, nickel, or other metal to enhance appearance or inhibit corrosion of the parent metal. Usually accomplished by immersion in an acid solution with cathode and anode electric current, causing the plating material to deposit on the parent metal.

plug weld: Weld made in a hole in one member of a lap joint, joining that member to that portion of the surface of the other member which is exposed through the hole. The walls of the hole may or may not be parallel, and the hole may be partially or completely filled with the weld metal.

poke welding: Spot welding process in which pressure is applied manually to one electrode. The other electrode is clamped to any part of the metal much in the same manner that arc welding is grounded.

polarity: In direct-current (DC) arc welding, tig welding, and wire feed welding, it's how current flows either positive to negative, or negative to positive. Alternating current (AC) welding has no polarity because it switches between positive and negative and back again 60 times per second.

porosity: Presence of gas pockets or inclusions in welding.

positions of welding: Welding is accomplished in one of four positions: flat, horizontal, overhead, and vertical. The limiting angles of the various positions depend somewhat as to whether the weld is a fillet or groove weld.

postheating: Application of heat to an assembly after a welding, brazing, soldering, thermal spraying, or cutting operation.

postweld interval: In resistance welding, the heat time between the end of weld time, or weld interval, and the start of hold time. During this interval, the weld is subjected to mechanical and heat treatment.

preheating: Application of heat to a base metal prior to a welding or cutting operation.

pressure controlled welding: Making spot or projection welds in which several electrodes function progressively under the control of a pressure sequencing device.

pressure welding: Welding process or method in which pressure is used to complete the weld.

preweld interval: In spot, projection, and upset welding, the time between the end of squeeze time and the start of weld time or weld interval during which the material is preheated. In flash welding, it is the time during which the material is preheated.

procedure qualification: Demonstration that welds made by a specific procedure can meet prescribed standards.

projection welding: Resistance welding process between two or more surfaces or between the ends of one member and the surface of another. The welds are localized at predetermined points or projections.

puddle: Liquid area of the weld where heat is being applied either by flame or electric arc. If the weld is properly controlled, the weld will be good.

pulsation welding: Spot, projection, or seam welding process in which the welding current is interrupted one or more times without the release of pressure or change of location of electrodes.

push welding: Making a spot or projection weld in which the corrected current is interrupted one or more times without the release of pressure or change of location of electrodes.

quenching: Sudden cooling of heated metal with oil, water, or compressed air.

reaction stress: Residual stress which could not otherwise exist if the members or parts being welded were isolated as free bodies without connection to other parts of the structure.

regulator: Device used to reduce cylinder pressure to a suitable torch working pressure.

reinforced weld: Weld metal built up above the surface of the two abutting sheets or plates in excess of that required for the size of the weld specified.

residual stress: Stress remaining in a structure or member as a result of thermal or mechanical treatment.

resistance brazing: Brazing process in which bonding is produced by the heat obtained from resistance to the flow of electric current in a circuit of which the work piece is a part, and by using a nonferrous filler metal having a melting point above 800 degrees F (427 degrees C), but below that of the base metals.

resistance butt welding: Resistance welding process in which the weld occurs simultaneously over the entire contact area of the parts being joined.

resistance welding: Welding process in which fusion is produced by heat obtained from resistance to the flow of electric current in a circuit of which the work piece is a part and by the application of pressure.

reverse polarity: Arrangement of direct current arc welding leads in which the work is the negative pole and the electrode is the positive pole of the welding arc.

Rockwell hardness test: In this test, a machine measures hardness by determining the depth of penetration of a penetrator into the specimen under certain arbitrary fixed conditions of test. The penetrator may be either a steel ball or a diamond spherocone.

root crack: Crack in the weld or base metal that occurs at the root of a weld.

root edge: Edge of a part to be welded which is adjacent to the root.

root face: Portion of the prepared edge of a member to be joined by a groove weld which is not beveled or grooved.

root of joint: Portion of a joint to be welded where the members approach closest to each other. In cross section, the root of a joint may be a point, a line, or an area.

root of a weld: Points at which the bottom of the weld intersects the base metal surfaces.

root opening: Separation between the members to be joined at the root of the joint.

root penetration: Depth a groove weld extends into the root of a joint measured on the centerline of the root cross section.

sandblasting: Method for cleaning certain metals before welding. A high velocity air blast, carrying sand, is directed at the metal and the particles of sand abrade its surface.

scarf: Chamfered surface of a joint.

scarfing: Process for removing defects that develop when steel billets are rolled, by using a low velocity oxygen deseaming torch.

seal weld: Weld used primarily to obtain tightness and to prevent leakage.

seam welding: Welding a lengthwise seam in sheet metal either by abutting or overlapping joints.

selective block sequence: Block sequence in which successive blocks are completed in a certain order selected to create a predetermined stress pattern.

series welding: Resistance welding process in which two or more welds are made simultaneously by a single welding transformer with the total current passing through each weld.

sheet separation: In spot, seam, and projection welding, the gap surrounding the weld between faying surfaces, after the joint has been welded.

shielded metal (stick) arc welding (SMAW): Arc welding process in which protection from the atmosphere is obtained through use of a flux, decomposition of the electrode covering, or an inert gas.

shielding gas: Gas that prevents contaminants from entering the molten weld pool.

shop weld: Prefabricating or welding subassemblies in a shop or controlled environment before taking them onsite for final assembly.

single impulse welding: Making spot, projection, and upset welds by a single impulse of current. When alternating current is used, an impulse may consist of a fraction of a cycle or a number of cycles.

slag: Oxidized impurities formed as a coating over the weld bead, or waste material found along the bottom edge of an oxyfuel cut.

slag inclusion: Non-metallic solid material entrapped in the weld metal or between the weld metal and the base metal.

slot weld: Weld made in an elongated hole in one member of a lap or tee joint joining that member to that portion of the surface of the other member which is exposed through the hole. The hole may be open at one end and may be partially or completely filled with weld metal.

slugging: Adding a separate piece or pieces of material in a joint before or during welding with a resultant welded joint that does not comply with design drawing or specification requirements.

soldering: Welding process that produces coalescence of materials by heating them to suitable temperature, and by using a filler metal having a liquidus not exceeding 450 degrees C (842 degrees F) and below the solidus of the base materials. The filler metal is distributed between the closely fitted surfaces of the joint by capillary action.

solidus: Highest temperature at which a metal or alloy is completely solid.

spacer strip: Metal strip or bar inserted in the root of a joint prepared for a groove weld to serve as a backing and to maintain the root opening during welding.

spall: Small chips or fragments which are sometimes given off by electrodes during the welding operation. This problem is especially common with heavy coated electrodes.

spatter: Metal particles expelled during arc and gas welding that do not form a part of the weld.

spot welding: Resistance welding process in which fusion is produced by the heat obtained from the resistance to the flow of electric current through the work pieces held together under pressure by electrodes. The size and shape of the individually formed welds are limited by the size and contour of the electrodes.

spray transfer: Type of metal transfer in which molten filler metal is propelled axially across the arc in small droplets.

staggered intermittent fillet weld: Two lines of intermittent welding on a joint, such as a tee joint, in which the fillet increments in one line are staggered with respect to those in the other line.

steel heat treating: Heating and rapidly cooling steel in the solid state to obtain certain desired properties, such as workability and corrosion resistance. Depending on the steel's mass and alloy type, it's oven-heated to 1475-to-1650 degrees F (802-to-900 degrees C), then quenched by dipping it in water or oil.

stick welding: Arc welding process in which protection from the atmosphere is obtained through use of a flux, decomposition of the electrode covering, or an inert gas.

stickout: Length of electrode (tungsten or wire) that sticks out past the gas lens, cup, or gun.

stitch weld: Tack-welding technique with short weld beads about ¾ inch long, spaced by equally long gaps with no welding. Used where a solid weld bead would be too costly and time-consuming, and where maximum strength is not required.

stored energy welding: Making a weld with electrical energy accumulated electrostatically, electromagnetically, or electrochemically at a relatively low rate and made available at the required welding rate.

straight polarity: Arrangement of direct current arc welding leads in which the work is the positive pole and the electrode is the negative pole of the welding arc.

stress relieving: Process of reducing internal residual stresses in a metal object by heating to a suitable temperature and holding for a proper time at that temperature. This treatment may he applied to relieve stresses induced by casting, quenching, normalizing, machining, cold working, or welding.

string bead welding: Method of metal arc welding on pieces ¾ inch (19 millimeters) thick or heavier in which the weld metal is deposited in layers composed of strings of beads applied directly to the face of the bevel.

stud welding: Arc welding process in which fusion is produced by heating with an electric arc drawn between a metal stud, or similar part, and the other workpiece, until the surfaces to be joined are properly heated. They are brought together under pressure.

submerged arc welding (SAW): Arc welding process in which fusion is produced by heating with an electric arc or arcs between a bare metal electrode or electrodes and the work. The welding is shielded by a blanket of granular, fusible material on the work. Pressure is not used. Filler metal is obtained from the electrode, and sometimes from a supplementary welding rod.

sugar: Crystallization in a weld. Usually occurs when welding stainless steel if the back side of the weld seam is not protected by an inert gas such as argon. Sugar has no strength and should be avoided in a weld.

surfacing: Depositing filler metal on a metal surface to obtain desired properties or dimensions.

tack weld: Weld made to hold parts of a weldment in proper alignment until the final welds are made.

tee joint: Joint between two members located approximately at right angles to each other in the form of a T.

temper colors: Colors which appear on the surface of steel heated at low temperature in an oxidizing atmosphere.

temper time: In resistance welding, the part of the postweld interval during which a current suitable for tempering or heat treatment flows. The current can be single or multiple impulse, with varying heat and cool intervals.

tempering: Reheating hardened steel to a temperature below the lower critical temperature, followed by a desired rate of cooling. The object is to release stresses set up, to restore some ductility, and to develop toughness through regulating or readjusting the structural constituents of the metal.

tensile strength: Maximum load per unit of original cross-sectional area sustained by a material during the tension test.

tension test: Test in which a specimen is broken by applying an increasing load to the two ends. During the test, the elastic properties and the ultimate tensile strength of the material are determined. After rupture, the broken specimen may be measured for elongation and reduction of area.

thermit crucible: Vessel in which the thermit reaction takes place.

thermit mixture: Mixture of metal oxide and finely divided aluminum with the addition of alloying metals as required.

thermit mold: A mold formed around the parts to be welded to receive the molten metal.

thermit reaction: Chemical reaction between metal oxide and aluminum which produces superheated molten metal and aluminum oxide slag.

thermit welding: Welding process in which fusion is produced by heating with superheated liquid metal and slag resulting from a chemical reaction between a metal oxide and aluminum, with or without the application of pressure. Filler metal, when used, is obtained from the liquid metal.

thoriated tungsten: Tungsten electrode with 1 or 2 percent thorium added to provide a more stable arc. Used to weld steel.

throat depth: In a resistance welding machine, the distance from the centerline of the electrodes or platens to the nearest point of interference for flatwork or sheets. In a seam welding machine with a universal head, the throat depth is measured with the machine arranged for transverse welding.

tig welding: Arc welding process in which fusion is produced by heating with an electric arc between a tungsten electrode and the work while an inert gas forms around the weld area to prevent oxidation. No flux is used.

toe crack: Crack in the base metal occurring at the toe of the weld.

toe of the weld: Junction between the face of the weld and the base metal.

torch brazing: Brazing process in which bonding is produced by heating with a gas flame and by using a nonferrous filler metal having a melting point above 800 degrees F (427 degrees C), but below that of the base metal. The filler metal is distributed in the joint of capillary attraction.

transverse seam welding: Making a seam weld in a direction essentially at right angles to the throat depth of a seam welding machine.

tungsten electrode: Non-filler metal electrode used in arc welding or cutting, made principally of tungsten.

ultrasonic testing: Process to test metal parts for defects. High-frequency sound waves are directed at the part, and their reflections are picked up by a receiver. Cracks and flaws inside the metal are detected by discontinuities in the return sound.

underbead crack: Crack in the heat-affected zone not extending to the surface of the base metal.

undercut: Groove melted into the base metal adjacent to the toe or root of a weld and left unfilled by weld metal.

undercutting: Undesirable crater at the edge of the weld caused by poor weaving technique or excessive welding speed.

upset: Localized increase in volume in the region of a weld, resulting from the application of pressure.

upset welding: Resistance welding process in which fusion is produced simultaneously over the entire area of abutting surfaces, or progressively along a joint, by the heat obtained from resistance to the flow of electric current through the area of contact of those surfaces. Pressure is applied before heating is started and is maintained throughout the heating period.

upsetting force: Force exerted at the welding surfaces in flash or upset welding.

vertical welding: Position of welding in which the axis of the weld is approximately vertical. In pipe welding, the pipe is in a vertical position and the welding is done in a horizontal position.

wandering block sequence: Block welding sequence in which successive weld blocks are completed at random after several starting blocks have been completed.

wandering sequence: Longitudinal sequence in which the weld bead increments are deposited at random.

wax pattern: Wax molded around the parts to be welded by a thermit welding process to the form desired for the completed weld.

weave bead: Type of weld bead made with transverse oscillation.

weaving: Technique of depositing weld metal in which the electrode is oscillated. It is usually accomplished by a semicircular motion of the arc to the right and left of the direction of welding. Weaving serves to increase the width of the deposit, decreases overlap, and assists in slag formation.

weld: Localized fusion of metals produced by heating to suitable temperatures. Pressure and/or filler metal may or may not be used. The filler material has a melting point approximately the same or below that of the base metals, but always above 800 degrees F (427 degrees C).

weld bead: Weld deposit resulting from a pass.

weld gauge: Device designed for checking the shape and size of welds.

weld metal: Portion of a weld that has been melted during welding.

weldability: Capacity of a material to form a strong bond of adherence under pressure or when solidifying from a liquid.

welding pressure: Pressure exerted during the welding operation on the parts being welded.

welding procedure: Detailed methods and practices including all joint welding procedures involved in the production of a weldment.

welding rod: Filler metal in wire or rod form, used in gas welding and brazing processes and in those arc welding processes in which the electrode does not provide the filler metal.

welding symbols: Symbols consisting of the following eight elements, or such of these as are necessary: reference line, arrow, basic weld symbols, dimension and other data, supplementary symbols, finish symbols, tail, specification, process, or other references.

welding technique: Details of a manual, machine, or semiautomatic welding operation which, within the limitations of the prescribed joint welding procedure, are controlled by the welder or welding operator.

welding tip: Tip of a gas torch especially adapted to welding.

welding torch: Device used in gas welding and torch brazing for mixing and controlling the flow of gases.

welding transformer: Device for providing current of the desired voltage.

weldment: Assembly whose component parts are formed by welding.

wire feed speed: Rate of speed in millimeters per second or inches per minute at which a filler metal is consumed in arc welding or thermal spraying.

work lead: Electric conductor (cable) between the source of arc welding current and the work piece.

yield point: Load per unit area value at which a marked increase in deformation of the specimen occurs with little or no increase of load; in other words, the yield point is the stress at which a marked increase in strain occurs with little or no increase in stress.

Index

• C •

• F •

• G •

• H •

• I •

• P •

• Q •

• R •

• S •

• T •

• U •

• V •

• W •

Making Everything Easier!
3rd Edition
Facebook
FOR
DUMMIES
Learn to:
Leah Pearlman
Carolyn Abram

Making Everything Easier!
Microsoft
Office 2010
ALL-IN-ONE
FOR
DUMMIES
8 BOOKS IN 1
Peter Weverka

Making Everything Easier!
iPad
FOR
DUMMIES
Learn to:
IN FULL COLOR!
Edward C. Baig
Bob "Dr. Mac" LeVitus

Making Everything Easier!
Windows 7
FOR
DUMMIES
Learn to:
Andy Rathbone

Dummies products make life easier!
DIY • Consumer Electronics • Crafts • Software • Cookware • Hobbies • Videos • Music • Games • and More!
For more information, go to **Dummies.com®** and search the store by category.

eMedia
Piano
For Dummies
The Fast and Easy Way To Start Playing!

The fun and easy way™ to pick up the basics and start expressing yourself!
Acrylic Painting Set
For Dummies
All you need to start painting right away!
An Art Set for the Rest of Us!™

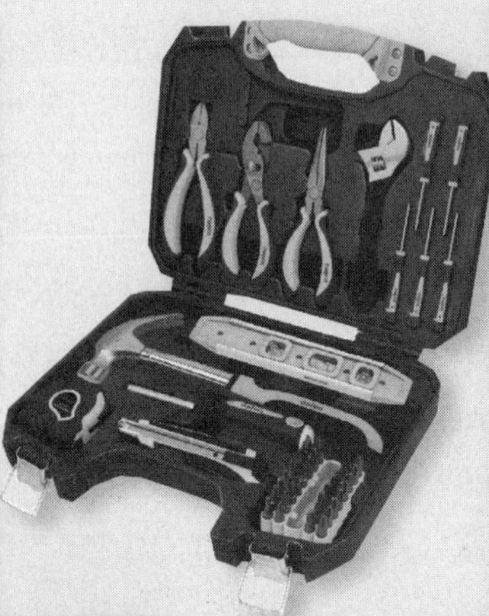

Drip Irrigation Kit
For Dummies

Plug In and Play — Electric Guitar the fun and easy way™!
Electric Guitar Starter Pack
For Dummies
Get ready to Rock!

DIY Project Calculator
Dummies

Windows 7 Transfer Kit
For Dummies
The easiest way to move to Windows 7

GPS Navigation
Dummies
MAYLONG

For Dummies®
Making everything easier!™